P9-DDL-077

Cody's Data Cleaning Techniques Using SAS®

Third Edition

Ron Cody

sas.com/books

The correct bibliographic citation for this manual is as follows: Cody, Ron. 2017. *Cody's Data Cleaning Techniques Using SAS®, Third Edition*. Cary, NC: SAS Institute Inc.

Cody's Data Cleaning Techniques Using SAS®, Third Edition

Copyright © 2017, SAS Institute Inc., Cary, NC, USA

ISBN 978-1-62960-796-2 (Hard copy)
ISBN 978-1-63526-067-0 (EPUB)
ISBN 978-1-63526-068-7 (MOBI)
ISBN 978-1-63526-069-4 (PDF)

All Rights Reserved. Produced in the United States of America.

For a hard copy book: No part of this publication may be reproduced, stored in a retrieval system, or transmitted, in any form or by any means, electronic, mechanical, photocopying, or otherwise, without the prior written permission of the publisher, SAS Institute Inc.

For a web download or e-book: Your use of this publication shall be governed by the terms established by the vendor at the time you acquire this publication.

The scanning, uploading, and distribution of this book via the Internet or any other means without the permission of the publisher is illegal and punishable by law. Please purchase only authorized electronic editions and do not participate in or encourage electronic piracy of copyrighted materials. Your support of others' rights is appreciated.

U.S. Government License Rights; Restricted Rights: The Software and its documentation is commercial computer software developed at private expense and is provided with RESTRICTED RIGHTS to the United States Government. Use, duplication, or disclosure of the Software by the United States Government is subject to the license terms of this Agreement pursuant to, as applicable, FAR 12.212, DFAR 227.7202-1(a), DFAR 227.7202-3(a), and DFAR 227.7202-4, and, to the extent required under U.S. federal law, the minimum restricted rights as set out in FAR 52.227-19 (DEC 2007). If FAR 52.227-19 is applicable, this provision serves as notice under clause (c) thereof and no other notice is required to be affixed to the Software or documentation. The Government's rights in Software and documentation shall be only those set forth in this Agreement.

SAS Institute Inc., SAS Campus Drive, Cary, NC 27513-2414

March 2017

SAS® and all other SAS Institute Inc. product or service names are registered trademarks or trademarks of SAS Institute Inc. in the USA and other countries. ® indicates USA registration.

Other brand and product names are trademarks of their respective companies.

SAS software may be provided with certain third-party software, including but not limited to open-source software, which is licensed under its applicable third-party software license agreement. For license information about third-party software distributed with SAS software, refer to **http://support.sas.com/thirdpartylicenses**.

Contents

List of Programs

Chapter 1 Working with Character Data

Chapter 2 Using Perl Regular Expressions to Detect Data Errors

Chapter 3 Standardizing Data

Chapter 4 Data Cleaning Techniques for Numeric Data

Chapter 5 Automatic Outlier Detection for Numeric Data

Chapter 6 More Advanced Techniques for Finding Errors in Numeric Data

Chapter 7 Describing Issues Related to Missing and Special Values (Such as 999)

Chapter 8 Working with SAS Dates

Chapter 9 Looking for Duplicates and Checking Data with Multiple Observations per Subject

Chapter 10 Working with Multiple Files

Chapter 11 Using PROC COMPARE to Perform Data Verification

Chapter 12 Correcting Errors

Chapter 13 Creating Integrity Constraints and Audit Trails

About This Book

What Does This Book Cover?

As the title implies, this is a book that shows you how to use SAS to identify (and fix) errors in your data. The book covers several different ways of detecting errors in character and numeric data.

There are several chapters devoted to checking character data, including simple DATA step programming, using formats to detect data errors, using Perl regular expressions to check that data values conform to a pre-determined pattern, and finally, standardizing such values as company names and addresses.

Checking for errors in numeric data is approached using several techniques. For some numeric values such as age or heart rate, you can use pre-determined ranges. For other types of numeric data, you can use techniques that automatically detect possible errors. This book also presents some statistical tests based on regression diagnostics that might find data errors when other methods fail.

This book includes a chapter describing several methods for correcting data errors and a chapter that describes SAS integrity constraints and audit trails. Briefly, integrity constraints are rules about your data that are stored inside a SAS data set that can block data errors when you add new data to an existing, previously cleaned data set.

Besides teaching you programming techniques, this book includes a collection of macros (pre-packaged SAS code) that you can use right out of the box or modify for your own particular task. Keep in mind that all of the programs, data, and macros in this book are available in a free download from the web site support.sas.com/cody.

Is This Book for You?

Just about anyone who is in the business of analyzing data needs to check that data for errors before engaging in any type of analysis. This book will save you the trouble of reinventing the wheel and having to create all your data cleaning programs from scratch.

What Are the Prerequisites for This Book?

This book assumes that you know some basic SAS programming skills. Those readers who are relatively new to SAS will appreciate the fact that all of the programs developed in the book are described in detail—those readers with more advanced programming skills will appreciate the innovative techniques developed to clean data.

What's New in This Edition?

This book is a third edition. Many second and third editions reflect minor changes from the original. This one does not. It has been over 10 years since the author wrote the second edition, and lots of things have changed since then—both the experience gained from teaching a data cleaning course for those 10 years and advances in SAS. This third edition is a major upgrade from its sibling. One of the first things you will notice are four new chapters, covering topics such as the use of Perl regular expressions for checking the format of character values (such as ZIP codes or email addresses) and how to standardize company names and addresses. One of the new chapters describes how to identify possible data errors in highly skewed distributions using regression diagnostics.

What Should You Know about the Examples?

Every program presented in this book is explained in detail, and you can run any of these programs yourself because they are all included in a free download of programs and data from the SAS author site.

Software Used to Develop the Book's Content

Most of the programs in this book can be run using base SAS, using any SAS platform that you typically use to analyze your data. A few of the more advanced techniques for detecting errors in numeric data use procedures found in SAS/STAT software.

SAS University Edition

 This book is compatible with SAS University Edition.

Output and Graphics

All of the output from the programs presented in this book are displayed with explanations and, in some cases, annotations.

We Want to Hear from You

SAS Press books are written *by* SAS users *for* SAS users. We welcome your participation in their development and your feedback on SAS Press books that you are using. Please visit https://support.sas.com/publishing to do the following:

- Sign up to review a book
- Recommend a topic
- Request information on how to become a SAS Press author
- Provide feedback on a book

Do you have questions about a SAS Press book that you are reading? Contact the author through saspress@sas.com or https://support.sas.com/author_feedback.

SAS has many resources to help you find answers and expand your knowledge. If you need additional help, see our list of resources: https://support.sas.com/publishing.

About The Author

 Ron Cody, EdD, a retired professor from the Rutgers Robert Wood Johnson Medical School, now works as a private consultant and a national instructor for SAS Institute Inc. A SAS user since 1977, Ron's extensive knowledge and innovative style have made him a popular presenter at local, regional, and national SAS conferences. He has authored or co-authored numerous books, as well as countless articles in medical and scientific journals.

Learn more about this author by visiting his author page at http://support.sas.com/cody. There you can download free book excerpts, access example code and data, read the latest reviews, get updates, and more.

Acknowledgments

I'm sitting here in my study, sipping a nice single malt, and relaxing a bit. Most of the creative work on my part is done—the book is out of my hands for a while.

Now is a good time to thank and acknowledge the wonderful group of people who, without whose help, I could not have written this book. First of all, a special expression of gratitude to my wife, Jan. The other day, she suggested that I was becoming obsessive. So, I want to thank her for making me get up every now and then and take a break, both physically and mentally. Speaking of family members, I also want to thank my son, Preston, for his help with regular expressions.

I offer a special thanks to Paul Grant, who not only went through every word and line of code I wrote, but compared sections of the third edition with corresponding sections of the second edition. Even I didn't do that! Paul has reviewed just about every book I have written for SAS and he keeps coming back for more. I also want to thank the two SAS employees, Charley Mullins and Kim Wilson, who reviewed the book.

This is the fourth book on which I have had the pleasure of working with Sian Roberts, as the acquisition and developmental editor. She really knows how to get things done—and quickly. Thanks, Sian.

Copy editing is much harder than it looks. I can read and re-read chapters multiple times and still miss mistakes. Not so with Kathy Restivo, my copy editor. She has been the copy editor for many of my books, and I am really grateful that she lets me write in my own style, while keeping me from embarrassment by correcting my spelling and grammar mistakes.

The last stage in putting the book together is performed by the production editor. This takes a lot of skill and I thank Monica McClain for her outstanding job.

Finally, the person who designs the cover. Once again, Robert Harris has come up with an eye-popping cover. Thanks, Robert.

Ron Cody

Introduction

What is data cleaning? Also referred to as data cleansing or data scrubbing, *data cleaning* has several definitions, depending on the type of data and the type of analysis that you are going to pursue. The basic idea is to analyze your data to look for possible errors. Data errors can be conveniently divided into two broad categories, depending on whether the data values are character or numeric.

As an example of a character error, consider a variable called Gender that has been defined as either an 'M' or an 'F'. A value of 'X' would clearly be a data error. What about a missing value for Gender? This could be considered a data error in a study where the gender of every subject needs to be known. What about an observation where a value of 'F' is coded, but the subject is a male? This is also a data error, but one that might be very difficult to detect. Other types of character data errors include misspelling names or other typographical errors. One other task involving character data consists of creating consistency of data across multiple observations or multiple data sets. For example, a company name might be coded as 'IBM' in one location and 'International Business Machines' in another location.

Numeric data errors are generally more straightforward. For some variables, such as pulse rate or blood pressure, data errors can often be identified by inspecting data values outside of a predetermined range. For example, a resting heart rate over 100 would be worthy of further investigation. For other types of numeric data, such as bank deposits or withdrawals, you could use automatic methods that look for what are called *outliers*—data values that are not consistent with other values.

The methods you use to search for bad data depend on several factors. If you are conducting a clinical trial for a drug application to the FDA, you need to ensure that all the values in your data set accurately reflect the values that were collected in the study. If a possible error is found, you need to go back to the original data source and attempt to find the correct value. If you are building a data warehouse for business applications and you find a data error, you might decide to throw out the offending data value.

SAS is especially well suited for data cleaning tasks. As a SAS programmer, you have extremely powerful character functions available in the DATA step. Many of the SAS procedures (PROCS) can be used as well.

As you can see in the Table of Contents, this book covers detecting character and numeric data errors. You will find chapters devoted to standardization of data and other chapters devoted to using Perl regular expressions to look for data values that do not conform to a specific pattern, such as US ZIP codes or phone numbers.

One of the later chapters discusses how to correct data errors and another chapter discusses how to create integrity constraints on a cleaned data set. Integrity constraints can automatically prevent data values that violate a predetermined definition from being added to the original clean data set. Along with integrity constraints, you will see how to create an audit trail data set that reports on data violations in the data you are attempting to add to the original clean data set.

Chapter 1: Working with Character Data

Introduction

This chapter will discuss and demonstrate some basic methods for checking the validity of character data. For some character variables such as Gender or Race, where there are a limited number of valid values (for example, 'M' and 'F' for Gender), it is quite straightforward to use PROC FREQ to list frequencies for these variables and inspect the output for invalid values.

In other circumstances, you might have rules that a set of character values must adhere to. For example, a variable such as ID might be stored as character, but there is a requirement that it must only contain digits. Other constraints on character data might be that all the characters are letters or, possibly, letters or digits (called *alphanumeric values*). SAS has some powerful tools to test rules of this sort.

This book uses several data sources where errors were added intentionally. One of these files, called Patients.txt, is a text file containing fictional data on 101 patients.

The data layout for this file is shown in Table 1.1:

Table 1.1: Data Layout for the File Patients.txt

Variable Name	Description	Starting Column	Length	Variable Type
Patno	Patient Number	1	3	Character
Account_No	Account Number	4	7	Character
Gender	Gender	11	1	Character
Visit	Visit Date	12	10	Numeric (SAS Date)
HR	Heart Rate	22	3	Numeric
SBP	Systolic Blood Pressure	25	3	Numeric
DBP	Diastolic Blood Pressure	28	3	Numeric
Dx	Diagnosis Code	31	7	Character
AE	Adverse Event	38	1	Numeric

The first few lines of Patients.txt is shown in Figure 1.1:

Figure 1.1: First Few Lines of the Patients.txt File

```
001CT14882M06/12/2012 69124 86713.4100
002MD78461M06/04/2010 76130 80047.5701
003DE 51381M06/22/2013 70 56 70108.5100
004CT37146M05/18/2013 76112 84669.8600
005DE00080F04/08/2012 91106 84078.1600
005DE00080F04/08/2012 91106 84078.1600
006DE37709M07/27/2014 71104 88967.5700
007VT56383F01/13/2014 63128 80640.2601
008PA67069F09/28/2013 79124 72020.1200
```

You can use the following SAS program to read this text file and create a SAS data set:

Program 1.1: Reading the Patients.txt File

```
libname Clean 'c:\books\Clean3';  ❶

data Patients;
   infile 'c:\books\clean3\Patients.txt' ;  ❷

   input  @1  Patno       $3.  ❸
          @4  Account_No $7.
          @11 Gender      $1.
          @12 Visit       mmddyy10.
          @22 HR          3.
          @25 SBP         3.
          @28 DBP         3.
          @31 Dx          $7.
          @38 AE          1.;
   label Patno      =     "Patient Number"  ❹
         Account_No = "Account Number"
         Gender     =     "Gender"
         Visit      =     "Visit Date"
         HR         =     "Heart Rate"
```

```
           SBP      =       "Systolic Blood Pressure"
           DBP      =       "Diastolic Blood Pressure"
           Dx       =       "Diagnosis Code"
           AE       =       "Adverse Event?";
      format Visit mmddyy10.;  ❺
run;

proc sort data=Clean.Patients;  ❻
   by Patno Visit;
run;

proc print data=Clean.Patients;  ❼
   id Patno;
run;
```

❶   The LIBNAME statement assigns the library reference Clean to the folder where the permanent data set will be stored (c:\books\Clean3).

❷   The text file Patients.txt is stored in the folder c:\books\Clean3. You use an INFILE statement to instruct the program to read data from the Patients.txt file in this folder.

❸   This type of input is called *formatted input*—you use a column pointer (@n) to indicate the starting column for each variable and an INFORMAT statement to instruct the program how to read the value. This INPUT statement reads character data with the n. informat, numeric data with the n. informat, and the visit date with the *mmddyy*10. informat.

❹   You use a LABEL statement to associate labels with variable names. These labels will appear when you run certain procedures (such as PROC MEANS) or in other procedures such as PROC PRINT (if you use a LABEL option). You do not have to use labels, but it makes some of the output easier to read.

❺   You use a FORMAT statement so that the visit date is in the *month-day-year* format (otherwise, it will appear as the number of days from January 1, 1960). You are free to use any date format you wish, such as date9. It does not have to be the same form that you used to read the date.

❻   You use PROC SORT to sort the data set by patient ID (Patno) and visit date (Visit).

❼   Use PROC PRINT to list the observations in the data set. The ID statement is an instruction to place this variable in the first column and to omit the Obs (observation number) column that PROC PRINT would include if you did not use an ID statement.

Before you look at the output from this program, take a look at the SAS Log to see if there were any errors in the program or in the data values. This is discussed in more detail in Chapter 7.

A listing of the first few lines is shown below:

Figure 1.2: Listing of the First Few Observations of Data Set Patients

Listing of data set Patients
Note: Data set sorted by Patno

Patno	Account_No	Gender	Visit	HR	SBP	DBP	Dx	AE
	DE56405	M	06/15/2010	87	128	98	195.920	0
001	CT14882	M	06/12/2012	69	124	86	713.410	0
002	MD78461	M	06/04/2010	76	130	80	047.570	1
003	DE51381	f	06/22/2013	70	56	70	108.510	0
004	CT37146	M	05/18/2013	76	112	84	669.860	0
005	DE00080	F	04/08/2012	91	106	84	078.160	0
005	DE00080	F	04/08/2012	91	106	84	078.160	0
006	DE37709	M	07/27/2014	71	104	88	967.570	0

Using PROC FREQ to Detect Character Variable Errors

We are going to start out by checking the character variables for possible invalid values. The variable Patno (stands for patient number) should only contain digits and should always have a value (for example, you do not want any missing values for this variable). The variable Account_No (account number) consists of a two-character state abbreviation followed by five digits. The variable Gender has valid values of 'M' and 'F' (uppercase). Missing values or values in lowercase are not allowed for Gender. Finally, the variable Dx (diagnosis code) consists of three digits, followed by a period, followed by three digits.

You can check Gender and the first two digits of the account number by computing frequencies, using the following program:

Program 1.2: Computing Frequencies Using PROC FREQ

```
libname Clean 'c:\Books\Clean3'; ❶

*Program to Compute Frequencies for Gender and the
 First Two Digits of the Account_No (State Abbreviation);

data Check_Char;
   set Clean.Patients(keep=Patno Account_No Gender); ❷
   length State $ 2; ❸
   State = Account_No; ❹
run;
```

```
title "Frequencies for Gender and the First Two Digits of Account_No";
proc freq data=Check_Char; ❺



    tables Gender State / nocum nopercent; ❻
run;
```

❶ The LIBNAME statement points to the folder c:\Books\Clean3, where the Patients data set is located. You can change this value to point to a valid location on your computer.

❷ The SET statement brings in the observations from the Patients data set. Note the use of the KEEP= data set option. Using the KEEP= data set option is more efficient than using a KEEP statement. This is an important point and, if you are not familiar with the distinction between a KEEP= data set option and a KEEP statement, pay close attention. If you had used a KEEP statement, all the variables from data set Patients would be imported into the PDV (program data vector – the place in memory that holds the variables and values). When you use a KEEP= data set option, only the variables Patno, Account_No, and Gender are read from data set Patients (and State is created in the program). For data sets with a large number of variables, using a KEEP= option is much more efficient then using a KEEP statement.

❸ The LENGTH statement sets the length of the new variable State to two bytes.

❹ Because State has a length of two bytes, assigning the value of Account_No to State results in State holding the first two characters of Account_No. You could have used the SUBSTR (substring) function to create the variable State, but the method used here is more efficient and used by many SAS programmers.

❺ Use PROC FREQ to compute frequencies.

❻ The TABLES statement lists the two variables (State and Gender) for which you want to compute frequencies. You can use the two options NOCUM and NOPERCENT to suppress the output of cumulative statistics and percentages.

Output from Program 1.2 is shown next:

Figure 1.3: First Part of the Output with Values for Gender

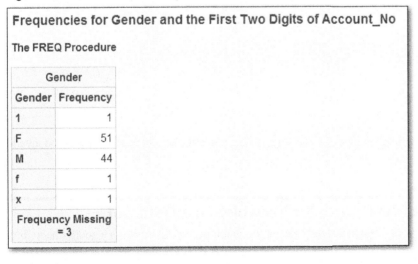

Frequencies for Gender and the First Two Digits of Account_No

The FREQ Procedure

Gender	
Gender	Frequency
1	1
F	51
M	44
f	1
x	1
Frequency Missing = 3	

There are several invalid values for Gender as well as three missing values.

The value of 'f' for Gender needs special attention. Raw data files sometimes contain values in mixed case. If you want to accept either upper- or lowercase values for Gender, you have several choices: The first option is to use the UPCASE function to change all lowercase values to uppercase. Another interesting and efficient option is to use the $UPCASE informat to convert all lowercase values for Gender (or any other variables you choose) to uppercase.

The following INPUT statement will convert all lowercase values for Gender to uppercase:

```
input   @1   Patno       $3.
        @4   Account_No $7.
        @11  Gender      $upcase1.
        @12  Visit       mmddyy10.
        @22  HR          3.
        @25  SBP         3.
        @28  DBP         3.
        @31  Dx          $7.
        @38  AE          1.;
```

Figure 1.4: Second Part of the Output with Values for State

State	Frequency
12	1
CT	15
DE	9
MA	8
MD	13
ME	8
NH	4
NJ	10
NY	11
PA	6
RI	6
VT	9
XX	1

There are several obviously invalid values for the state codes. You will see which other values are invalid later in this chapter.

Changing the Case of All Character Variables in a Data Set

While we are on the topic of ensuring consistent casing for character variables, this is a good time to discuss a program that will change the case for all of the character variables in a SAS data set. You can create an array of all the character variables in a data set by using the keyword _CHARACTER_ in the ARRAY statement. Once the array is created, you can perform operations on all of the character variables. You only have a handful of character variables in the Patients data set, but this data set can be used to demonstrate this technique. Here it is:

Program 1.3: Programming Technique to Perform an Operation on All Character Variables in a Data Set

```
*Program to UPCASE all the character variables in the
 Patients data set;

data Clean.Patients_Caps;
   set Clean.Patients;
   array Chars[*] _character_; ❶
   do i = 1 to dim(Chars); ❷
      Chars[i] = upcase(Chars[i]); ❸
   end;
   drop i;
run;

title "Listing the First 10 Observations in Data Set Patients_Caps";
proc print data=clean.Patients_Caps(obs=10) noobs; ❹
run;
```

❶ Use the keyword _CHARACTER_ to create an array of all the character variables in data set Patients. It is important to note that the keyword _CHARACTER_, when used in a DATA step, refers to all the character variables **at that point in the DATA step**. This is important, because had you placed the ARRAY statement before the SET statement, there would be no variables in the array.

❷ Because you do not always know how many character variables are in the data set you want to process, you use an asterisk (*) instead of the actual number. The DIM function takes as its argument the name of an array and returns the number of elements (variables) in the array.

❸ Each element of the array is converted to uppercase. You could substitute LOWCASE or PROPCASE (proper case) functions instead of UPCASE if you wish.

❹ The data set option OBS= is used to print the first 10 observations in the Patients_Caps data set.

Note: The LIBNAME statement defining the Clean library will no longer be included in the sample programs: The LIBNAME statement defining the Clean library was copied to the AUTOEXEC.SAS file so that it is automatically assigned every time a SAS session is started.

The listing is shown below: Note that the lowercase value for Gender in patient 003 is now in uppercase.

Figure 1.5: Partial Listing of Data Set Patients_Caps

Listing the First 10 Observations in Data Set Patients_Caps

Patno	Account_No	Gender	Visit	HR	SBP	DBP	Dx	AE
	DE56405	M	06/15/2010	87	128	98	195.920	0
001	CT14882	M	06/12/2012	69	124	86	713.410	0
002	MD78461	M	06/04/2010	76	130	80	047.570	1
003	DE51381	F	06/22/2013	70	56	70	108.510	0
004	CT37146	M	05/18/2013	76	112	84	669.860	0
005	DE00080	F	04/08/2012	91	106	84	078.160	0
005	DE00080	F	04/08/2012	91	106	84	078.160	0
006	DE37709	M	07/2044	71	104	88	967.570	0

A Summary of Some Character Functions (Useful for Data Cleaning)

This section is a review of some of the SAS character functions that are useful for data cleaning applications.

UPCASE, LOWCASE, and PROPCASE

These three functions change the case of character values. You have already seen the UPCASE function in Program 1.3. The LOWCASE function is similar except that this function converts all character values to lowercase. It is worthwhile to take a minute to review the PROPCASE function. Proper case capitalizes the first letter of every word and converts the remaining characters to lowercase. For example, PROPCASE("roN cODY") returns the value "Ron Cody." PROPCASE has an optional second argument that you can use to specify word delimiters. The default delimiter is a blank. However, it is often useful to add a single quote as an alternate delimiter. An example that shows the utility of adding a single quote as a delimiter is PROCPCASE("d'amore"). The result is "D'amore." If you include a single quote as an additional delimiter like this PROPCASE("d'amore"," '"), the result is "D'Amore."

NOTDIGIT, NOTALPHA, and NOTALNUM

These three functions search a character value and return the position of the first character in that value that is not a digit (NOTDIGIT), the first character that is not a letter (NOTALPHA), or the first character that is not an alphanumeric value (NOTALNUM). If the search fails, the functions return a 0. These three functions have an optional second argument to specify the starting position in a string to begin the search. If you make the starting position negative, the search proceeds from right to left, starting at the position corresponding to the absolute value of the starting position. Here are some examples:

```
notdigit("123x456") = 4 (the position of the 'x')
notdigit("12345") = 0 (there are no non-digits)
notdigit("NY12345") = 1 (the position of the 'N')
notdigit("NY12345",3) = 0 (the search is starting from position 3)
notalpha("NY12345") = 3 (the position of the '1')
notalpha("Oscar") = 0 (there are no non-alpha characters)
```

```
notalnum("Ron Cody") = 4 (the position of the blank)
notalnum("abc123") = 0 (there are no non-alphanumeric values)
```

VERIFY

The VERIFY function is useful when the rules concerning a character value are not as clear-cut as the rules that you can test with NOTDIGIT, NOTALPHA, or NOALNUM. This function allows you to specify which characters to consider valid. The first argument to the VERIFY function is the character value you want to check—the second argument is the list of valid characters (known as the *verify list*). If all the values in the character string that you are testing are found in the verify list, the function returns a 0 (this is usually a good thing—it means there were no bad characters in the string). If there are one or more characters in the string you are testing that are not in the verify list, the function returns the position of the first bad character. Here are some examples:

```
verify("X123","12345XYZ") = 0 (the characters 'X', '1', '2', and '3' are all in the verify string)
verify("X4321","123XYZ") = 2 ('4' is not in the verify string)
verify("12345","0123456789") = 0 (the same as notdigit("12345"))
```

COMPBL

The COMPBL function (stands for compress blanks) converts all multiple blanks to a single blank. This is a useful function if you have data that might contain more than one blank between words. Here are some examples:

```
compbl("Fred     Flintstone") = "Fred Flintstone"
compbl("Ron   Cody") = "Ron Cody"
compbl("12 New   Town   Road" = "12 New Town Road"
```

COMPRESS

This is an amazing function that belongs in my list of the top 10 SAS functions. The COMPRESS function can remove specified characters from a string. If you use the 'k' modifier to specify which characters you want to keep, the function will keep all the specified characters and throw away all the others.

The COMPRESS function can take three arguments. The first argument is the character value you plan to modify. If you supply only one argument, the default action removes blanks from the first argument. The second argument is a list of characters that you want to remove from the string (unless you use the 'k' modifier). If you provide only two arguments to the COMPRESS function, it will remove all the characters you specify in the second argument. The third argument has several purposes: First, you can use modifiers to select a class of characters to remove (or keep). Some of the more useful modifiers are:

a

　　upper- and lowercase letters

d

　　digits

s

　　space characters (blank, tabs, line feeds, carriage returns)

p

　　punctuation

i

    ignores case

k

    keep

If you include any of the first four modifiers in this list, the COMPRESS function will remove all the characters you list in the second argument plus all the characters specified by the modifiers. However, if you include the 'k' modifier, the function does not remove the specified characters—it keeps the specified characters and removes everything else. Some examples are in order:

compress("Abe Lincoln") = "AbeLincoln" (with one argument, COMPRESS removes blanks)

compress("ABC123","B2") = "AC13" (removes 'B' and '2' from the string)

compress("abc12345xyz",,"a") = "12345" (removes upper- and lowercase letters)

Notice the two commas in the parentheses. Why are they needed? With a single comma, the function would treat the 'a' as the second argument—the two commas indicate that the 'a' is the third argument.

compress("10 kgs.",,"kd") = "10" (keeps the digits, throws away everything else)

compress("AaBbCc","ab","i") = "Cc" (removes 'a' and 'b', ignoring case)

compress("12345abcde","3","a") = "1245" (removes '3' plus all upper- and lowercase letters)

MISSING

This is my favorite SAS function. It tests if the argument is a character or numeric missing value. If the argument is a missing value, the function returns a value of true—if the argument is not a missing value, the function returns a value of false. The reason I like this function so much is that it makes your program much easier to read. For example, without the MISSING function, you might write lines of code like this:

```
if Age = . then put "Age is missing";
if Name = ' ' then put "Missing Name";
```

With the MISSING function, these two statements become:

```
if missing(Age) then put "Age is missing";
if missing(Name) then put "Missing Name";
```

TRIMN and STRIP

The TRIMN function removes trailing blanks from a string. (Note: The TRIMN function is similar to the older TRIM function except that TRIM returns a single blank if the argument is a missing value—the TRIMN function returns a null string in this situation.)

The STRIP function removes leading and trailing blanks from a string. Here are some examples of both functions:

```
trimn("Hello   ") || "X" = "HelloX"
":" || strip("   abc   ") || ":" = :abc:
```

You will see all of these functions used in data cleaning tasks throughout this book.

Checking that a Character Value Conforms to a Pattern

Because the Dx variable has many unique values, PROC FREQ will produce a long list that will be hard to inspect manually. If you had a list of valid Dx codes, you could write a program to check for invalid Dx codes. Without such a list, you can still do a limited check on the Dx variable. In this example, Dx codes, as mentioned previously, consist of three digits, a period, followed by three more digits.

Note: These are not ICD-9 or ICD-10 codes, commonly used to code medical diagnoses. ICD codes can also contain letters. These Dx codes are "Ron's Codes."

You can write a program to ensure that all the Dx codes are consistent with this pattern. Program 1.4 is such a program:

Program 1.4: Checking that the Dx Values Are Consistent with the Required Pattern

```
title 'Checking for Invalid Dx Codes';

data _null_;
   set Clean.Patients(keep=Patno Dx);   ❶
   length First_Three Last_Three $ 3 Period $ 1; ❷
   First_Three = Dx;   ❸
   Period = substr(Dx,4,1);  ❹
   Last_Three = substr(Dx,5,3);  ❺
   file print;  ❻
   if missing(Dx) then put
      "Missing Dx for patient " Patno;  ❼
   else if notdigit(First_Three) or Period ne '.' or notdigit(Last_Three)
      then put "Invalid Dx " Dx "for patient " Patno;
run;
```

This DATA step uses the reserved data set name _NULL_. If you are not familiar with DATA _NULL_, here is a short explanation:

Because the only purpose of this program is to identify invalid data values and print them out, there is no need to create a SAS data set. The reserved data set name _NULL_ tells SAS not to create a data set. This saves all the overhead of creating a real data set and writing information to it—a major efficiency technique. When the DATA step completes, there is no data set. Therefore, you typically use PUT statements to either print messages to the log or output destination or, possibly, to a disk file. The idea is that if you don't need a data set, don't create one.

❶ The SET statement brings in the two variables Patno and Dx from the Patients data set.
❷ The LENGTH statement sets the lengths of the three character variables you are going to extract from the three parts of Dx (the first three characters, the period, and the last three characters).
❸ This statement uses the same trick used in Program 1.2: Because the length of First_Three is three bytes (due to the LENGTH statement), this assignment statement places the first three characters of Dx and assigns them to the variable First_Three.
❹ The SUBSTR function extracts a substring from the first argument, starting at the position of the second argument, for a length specified by the third argument. This statement assigns the fourth character of Dx to the variable called Period.
❺ You use the SUBSTR function again to assign the last three characters of Dx to the variable Last_Three.

❻ This FILE statement sends the data to the output device (PRINT is a reserved keyword). If you leave off this statement, the default destination of a PUT statement is the SAS Log.

❼ The next two statements write error messages if the value of Dx is a missing value or if the pattern doesn't match the pattern required for the variable Dx.

Running Program 1.4 results in the following output:

Figure 1.6: Values of Dx that Do Not Meet the Required Pattern

```
Invalid Dx 530.abc for patient 011
Missing Dx for patient 091
Invalid Dx V23.000 for patient 094
```

You see three Dx codes that are either missing or that do not meet the required pattern. In the next chapter on Perl regular expressions, you will see how a regular expression can be used to test for specified patterns in character data.

Using a DATA Step to Detect Character Data Errors

Earlier in this chapter you saw how to use PROC FREQ to detect invalid values for Gender and State codes. While this is a useful first step, it doesn't identify the patient numbers for the invalid values. Just as you used a DATA step to detect invalid Dx codes, you can use a similar method with Gender and State.

Program 1.5: Using a DATA Step to Identify Invalid Values for Gender and State Codes

```
*Using a DATA step to check for invalid Gender and State Codes;

data _null_;
   title "Invalid Gender or State Codes";
   title2;
   file print;
   set Clean.Patients(keep=Patno Gender Account_No);
   length State $ 2;
   State = Account_No;

   *Checking value of Gender;
   if missing(Gender) then put
      "Patient " Patno "has a missing value for Gender";
   else if Gender not in ('M','F') then put
      "Patient number " Patno "has an invalid value for Gender: " Gender;

   *Checking for invalid State abbreviations;
   if State not in ('NJ','NY','PA','CT','DE','VT','NH','ME','RI','MA','MD')
      then put "Patient number " Patno "has an invalid State code: " State;
run;
```

You first check for missing or invalid values for Gender and print appropriate error messages. You can use the IN operator to check if your state abbreviation is among the 11 valid state codes (most of the New England states). Here is the output:

Figure 1.7: Output from Program 1.5

```
Invalid Gender or State Codes

Patient number 008 has an invalid value for Gender: f
Patient 027 has a missing value for Gender
Patient number 039 has an invalid State code: 12
Patient number 041 has an invalid State code: xx
Patient 055 has a missing value for Gender
Patient 058 has a missing value for Gender
Patient number 088 has an invalid value for Gender: x
Patient number 095 has an invalid value for Gender: 1
```

Using PROC PRINT with a WHERE Statement to Identify Data Errors

You can use PROC PRINT with a WHERE statement to perform some limited data checking. This method is especially useful for a quick check on a limited number of variables, especially for individuals who have limited programming experience. As an example, you can identify missing or invalid patient numbers and genders with the following program:

Program 1.6: Using PROC PRINT with a WHERE Statement to Check for Data Errors

```
*Using PROC PRINT to identify data errors;

title "Using PROC Print to Identify Data Errors";
proc print data=Clean.Patients;
    id Patno;
    var Account_No Gender;
    where notdigit(Patno) or
          notalpha(Account_No,-2) or
          notdigit(Account_No,3) or
          Gender not in ('M','F');
run;
```

You use the NOTDIGIT function to ensure that the variable Patno only contains digits. The first condition testing the variable Account_No uses the NOTALPHA function with a starting position of -2. The search starts at position 2 (the absolute value of the starting position) and the search proceeds from right to left because of the minus sign. Although this condition does not identify invalid state abbreviations, it does identify any account number that has a non-alpha character in the first two positions. You use the NOTDIGIT function with a starting position of 3 to check the account number for non-digits, starting from position 3. Here are the results:

Figure 1.8: Using PROC PRINT with a WHERE Statement to Check for Data Errors

Using PROC Print to Identify Data Errors

Patno	Account_No	Gender
	DE56405	M
003	DE51381	f
027	MD40964	
039	1234567	F
055	MD08716	
058	NY1234z	
088	PA50872	x
095	DE44197	1
XX5	MA93350	F

Here you see a missing patient number and an invalid patient number (XX5). The account number for patient 039 has non-alpha characters in the first two positions. All of the invalid values for Gender are listed as well. It is important to realize that some of the observations in this listing might contain multiple errors. For example, patient 058 has an invalid account number and a missing Gender.

Using Formats to Check for Invalid Values

Another way to check for invalid values of a character variable is with user-defined formats. Let's use a user-defined format to check for invalid values of Gender. Program 1.7 formats all values of Gender to 'Valid', 'Missing', or 'Error'. It then uses PROC FREQ to compute frequencies for each of these three values.

Program 1.7: Using a User-Defined Format to Check for Invalid Values of Gender

```
*Using formats to identify data errors;

title "Listing Invalid Values of Gender";
proc format;
   value $Gender_Check 'M','F' = 'Valid'
                       ' '      = 'Missing'
                       other    = 'Error';
run;

proc freq data=Clean.Patients;
   tables Gender / nocum nopercent missing;
   format Gender $Gender_Check.;
run;
```

Besides the two TABLES options NOCUM and NOPERCENT, this program includes the MISSING option as well. The MISSING option treats missing values as a valid category and places counts for missing values in the body of the table instead of at the bottom of the listing.

> **Important note:** When you use the MISSING TABLES option with PROC FREQ and you are outputting percentages, the percentages are computed by dividing the frequency of a particular value by the total number of observations, missing or non-missing.

Because you included a FORMAT statement in PROC FREQ, it will compute frequencies on the formatted values. Here is the output:

Figure 1.9: Output from PROC FREQ

Listing Invalid Values of Gender

The FREQ Procedure

Gender	
Gender	Frequency
Missing	3
Error	3
Valid	95

There are 3 missing values, 3 errors, and 95 valid values of Gender in the Patients data set. While this is useful information, the next program uses the user-defined format in a DATA step to tell you which patients have invalid (or missing) values of Gender.

Program 1.8: Using a User-Defined Format to Identify Invalid Values

```
*Using formats to identify data errors;

title "Listing Invalid Values of Gender";
proc format;
   value $Gender_Check 'M','F' = 'Valid'
                       ' '     = 'Missing'
                       other   = 'Error';
run;

data _null_;
   set Clean.Patients(keep=Patno Gender);
   file print;
   if put(Gender,$Gender_Check.) = 'Missing' then put
      "Missing value for Gender for patient " Patno;
   else if put(Gender,$Gender_Check.) = 'Error' then put
      "Invalid value of " Gender "for Gender for patient " Patno;
run;
```

Before we show output from this program, let's review the PUT function. It is useful to first think about what a PUT statement does. It takes a value (character or numeric) and writes out formatted values to a file or other location. The PUT function performs a similar operation. It takes a value, formats this value, and "writes" out the value to a character variable. Because format values are always character values, the result of a PUT function is always a character value. To help clarify this, take a look at the following SAS statement:

```
     Char_Var = put(Gender,$Gender_Check.);
```

Char_Var will have one of three values: 'Valid', 'Missing', or 'Error'. By the way, the length of a variable created with a PUT function is the length of the longest formatted value. In this example, the length of Char_Var is 7 (the length of the value 'Missing').

Now that you understand how the PUT function works, it's time to run the program and examine the output. Here it is:

Figure 1.10: Output from Program 1.8

```
Listing Invalid Values of Gender

Invalid value of f for Gender for patient 003
Missing value for Gender for patient 027
Missing value for Gender for patient 055
Missing value for Gender for patient 058
Invalid value of x for Gender for patient 088
Invalid value of 1 for Gender for patient 095
```

Why would you use formats to check for invalid character values when this was easily done without all the added complication? There are several advantages to the format approach. First, if you have multiple studies with similar variables, you can create your formats and place them in a permanent format library. You can then use one or two SAS statements (using the PUT function) to test for invalid values. Another advantage to using formats is efficiency. Formats are stored in memory so if you have very large data sets, a program using formats should run faster than an alternative program that does not use formats.

Creating Permanent Formats

You can make the use of user-defined formats for data checking more powerful and efficient if you create and use permanent formats. Some SAS programmers believe this is a daunting process attempted only by veteran programmers. Not true. Just follow the steps presented here to create permanent formats and you will become a power programmer.

First, decide where you want to store your formats. You can store them in the same library that you use to store your permanent data sets or you can create a library just for formats. In this example, you are going to place your permanent formats in the library called Clean, the one where the Patients data set and several others are stored. The SAS statement to create this library was:

```
     libname Clean 'c:\books\Clean3';
```

Of course, you will be using a different folder on your machine.

Next, you want to add a statement to your AUTOEXEC.SAS program, the one that automatically executes every time you begin a SAS session. Here is the line you need to add:

```
     options fmtsearch=(Clean);
```

This instruction tells SAS to look for your formats in the Clean library. Remember that you must execute your LIBNAME statement before the FMTSEARCH= option.

Next, add the option `LIBRARY=Clean` to PROC FORMAT. For example, rerunning the PROC FORMAT statements that created the $Gender format to make this format permanent look like this:

Program 1.9: Making a Permanent Format

```
proc format library=Clean;
   value $Gender_Check 'M','F' = 'Valid'
                       ' '     = 'Missing'
                       other   = 'Error';
run;
```

Now, run PROC FORMAT (just once). Every time you start a SAS session, the FMTSEARCH= option will execute (because the statement was added to the AUTOEXEC.SAS file) and your user-defined formats will be available for use in the same way you would use SAS formats such as *mmddyy*10. or $dollar10.2.

Removing Units from a Value

It is not uncommon to have numeric data that contains units such as Kgs. or Lbs. The following list of values can serve as an example:

Table 1.2: Weights with Units Attached

```
100lbs.
110 Lbs.
50Kgs.
70 kg
180
```

Given these weights, you want to create a numeric variable that represents weight in kilograms. Thanks to the COMPRESS function, this is an easy task.

Program 1.10: Converting Weight with Units to Weight in Kilograms

```
*Program to Remove Units from Numeric Data;

data Units;
   input Weight $ 10.;
   Digits = compress(Weight,,'kd'); ❶
   if findc(Weight,'k','i') then   ❷
      Wt_Kg = input(Digits,5.);
   else if not missing(Digits) then
      Wt_Kg = input(Digits,5.)/2.2; ❸
datalines;
100lbs.
110 Lbs.
50Kgs.
70 kg
180
;
title "Reading Weight Values with Units";
proc print data=Units noobs;
   format Wt_Kg 5.1;
run;
```

You might not be familiar with the DATALINES statement. Instead of writing an external file and using an INFILE statement, you can place your data right after the DATALINES statement. The INPUT statement will read the values as if they were in an external file.

❶ Using the two modifiers, 'k' and 'd' (**k**eep the **d**igits), extract the digits from the string, throwing away everything else. Note that the variable Digits is a character variable.

❷ Use the FINDC function to determine if there is an upper- or lowercase 'k' in the original Weight variable. The FINDC function searches the first argument for any of the characters listed in the second argument. The third argument (the 'i') is a modifier that says to ignore case. Thus, the expression will be true if there is an upper- or lowercase 'k' in the Weight value. If a 'k' is found, the INPUT function performs a character-to-numeric conversion.

❸ If there is no 'k' in the Weight value and Weight is not missing, perform the character-to-numeric conversion using the INPUT function and divide the result by 2.2 to convert pounds to kilograms. Note that it is assumed that any Weight value that does not contain a 'k' is in pounds. The output is listed next:

Figure 1.11: Output from Program 1.10

Reading Weight Values with Units

Weight	Digits	Wt_Kg
100lbs.	100	45.5
110 Lbs.	110	50.0
50Kgs.	50	50.0
70 kg	70	70.0
180	180	81.8

Removing Non-Printing Characters from a Character Value

You might have come across data, either in ASCII, EBCDIC, or Unicode (a superset of ASCII), that includes codes that do not correspond to any valid character. You might want to remove these values from your data. Once again, the COMPRESS function comes to the rescue.

If you only want to keep letters and digits, you could use the following expression:

```
Char_Value = compress(Char_Value,,'kad');
```

The expression to keep all letters, digits, punctuation, and space characters (this includes tabs) is:

```
Char_Value = compress(Char_Value,,'kadps');
```

How about a more complicated scenario? What if you want to keep letters, digits, periods, spaces, and minus signs, but you do not want other white space characters or punctuation. To accomplish this, you need to provide a second and third argument, like this:

```
Char_Value = Char_Value,'. -','kad');
```

The second argument specifies three characters: a period, space, and a minus sign. The third argument includes the 'k' modifier (keep the characters listed) as well as the 'a' modifier (all upper- and lowercase letters) and the 'd' modifier (all digits). You can use these examples as a guide to using the COMPRESS function for any similar situation.

Conclusions

Most of the techniques for checking the validity of character data are straightforward. For variables with only a few possible valid values (such as Gender or Race), PROC FREQ can determine if there are invalid values in the data set. You can use a DATA _NULL_ program to identify which subjects have invalid values. In some cases, PROC PRINT followed by a WHERE statement can identify errors. Another way to identify invalid character data is to create a user-defined format specifying what values are valid, invalid, or missing. Finally, the COMPRESS function is useful in removing units from a value or in removing non-printing characters from data.

Chapter 2: Using Perl Regular Expressions to Detect Data Errors

Introduction

Regular expressions were first introduced in a language called Perl, developed for UNIX and Linux systems. Perl regular expressions were added to SAS starting with SAS 9.0. They are used to describe text patterns. For example, you can write an expression that matches a Social Security number (three digits, a dash, two digits, a dash, and four digits). Therefore, you can use a regular expression to test that a specific pattern is present or not. Looking back at Chapter 1, you wanted to ensure that the diagnosis codes consisted of three digits, a period, followed by three more digits. While it was possible to do this using SAS character functions, using a regular expression is much simpler.

Describing the Syntax of Regular Expressions

There are entire books devoted to regular expressions. The goal of this chapter is to describe some basic aspects of regular expressions and provide some examples of how they can be used.

A regular expression (called a *regex* by programmers) starts with a delimiter (most often a forward slash), followed by a combination of characters and meta-characters. *Meta-characters* describe character classes such as all digits or all punctuation marks. The expression ends with the same delimiter you started with. For example, a regular expression for a Social Security number is:

```
/\d\d\d-\d\d-\d\d\d\d/
```

\d is the meta-character for any digit. The two dashes in the expression are just that—dashes. Any character that is not defined as a special character, such as the dashes in the expression above, represent themselves. Even spaces count as characters in a regular expression.

Because writing \d four times is tedious, this expression can be rewritten as:

```
/\d\d\d-\d\d-\d{4}/
```

You can probably guess that the {4} following the \d says to repeat \d four times.

You can create sets of characters using square brackets [and]. All the uppercase letters are represented by [A-Z]. All upper- and lowercase letters are represented by [A-Za-z].

Likewise, the digits 0-9 can be represented by [0-9]. If you are using ASCII, you can also use \d instead of [0-9].

In SAS, all the regular expression functions begin with the letters PRX (**P**erl **R**egular **Ex**pression). Let's write a program to check that the Dx code in the Patients data set satisfies the required pattern.

Program 2.1: Using a Regex to Test the Dx Values

```
title "Checking Dx Values Using a Regular Expression";

data _null_;
   file print;
   set Clean.Patients(keep=Patno Dx);
   if not prxmatch("/\d\d\d\.\d\d\d/",Dx) then
      put "Error for patient " Patno "  Dx code = " Dx;
run;
```

The PRXMATCH function takes two arguments. The first argument is a regular expression—the second argument is the string you are examining. If the string contains a pattern described by the regular expression, the function returns the starting position for the pattern. If the pattern described by the regular expression is not found in the string, the function returns a 0. In Program 2.1, the regular expression is describing a pattern of three digits, a period, followed by three digits.

Notice the backslash (\) before the period in the regular expression above. As mentioned previously, certain characters in a regular expression have special meaning. A period is one of those special characters—it stands for any character (sometimes referred to as a *wildcard*). In order to specify that you are looking for a period and not a placeholder for a character, you place the backslash character before the period. The backslash, called an *escape character* by regex experts, says to treat the following character as a character and not the special meaning usually described by that character.

When a non-valid Dx code is tested, the pattern is not found and the function returns a 0. When a valid pattern is found, the function returns a 1 (because all Dx codes start in the beginning of the string). Therefore, the error message is written out for all Dx codes that do not match the pattern. Here is the output:

Figure 2.1: Output from Program 2.1

Checking Dx Values Using a Regular Expression

```
Error for patient 011    Dx code = 530.abc
Error for patient 091    Dx code =
Error for patient 094    Dx code = V23.000
```

The same three observations that you identified in Chapter 1 are listed here.

Checking for Valid ZIP Codes and Canadian Postal Codes

You can use a program similar to Program 2.1 to verify that an address contains a valid ZIP code, either a five-digit code or a five-digit code followed by a dash, followed by four more digits (ZIP code +4).

A regular expression to check either the five-digit code or the ZIP+4 code is:

```
/\d{5}(-\d{4})?/
```

The first part of this expression is pretty clear—it says to search for five digits. The expression in parentheses matches a dash followed by four digits. The question mark following the parentheses means to search for 0 or 1 occurrences of the previous expression. Therefore, this expression matches either of the two valid US ZIP code formats.

You can run the following program to test this expression:

Program 2.2: Testing the Regular Expression for US ZIP Codes

```
*Program to test the Regex for US Zip Codes;

title "Testing the Regular Expression for US Zip Codes";
data _null_;
   file print;
   input Zip $10.;
  if not prxmatch("/\d{5}(-\d{4})?/",Zip) then
      put "Invalid Zip Code " Zip;
datalines;
12345
78010-5049
12Z44
ABCDE
08822
;
```

This program reads in the ZIP code (allowing for up to 10 characters) and prints out an error message for any code that does not match the regular expression. Here is the output from Program 2.2:

Figure 2.2: Output from Program 2.2

```
Testing the Regular Expression for US Zip Codes

Invalid Zip Code 12Z44
Invalid Zip Code ABCDE
```

Canadian postal codes are a bit more complicated. A valid Canadian postal code takes one of two possible forms, as follows:

LdLdLd ***or*** LdL dLd

where 'L' is a letter and 'd' is a digit. Not all the letters of the alphabet are allowed. The letters D, F, I, O, Q, and U are never used (the forbidden six). In addition, the first letter of the postal code also cannot be a W or a Z. Writing a regular expression for Canadian postal codes is, therefore, a bit trickier. The simplest expression

/\[A-Z]\d\[A-Z]\d[A-Z] ?\d[A-Z]\d/

ignores the forbidden letters but can be used to verify the basic form of the code. The blank space and question mark in the expression allow for either no space or one space in the middle of the code.

A more exact expression is:

/[ABCEFGHJ-NPQRSTVXY][0-9][ABCEFGHJ-NPRSTV-Z] ?[0-9]
[ABCEFGHJ-NPRSTV-Z][0-9]/

There are a few features of this expression that need some explanation. First, the list of letters in the square brackets means that any one of those characters is valid in that position. The first list of letters does not include the six forbidden characters or W or Z. Instead of using \d to represent any digit, the expression [0-9] is used. The difference between \d and [0-9] only presents itself if you have a file that is coded in UNICODE rather than ASCII. It turns out that with non-ASCII coding, there are characters that match \d that are not what you would consider digits. The other two lists of letters include all the letters of the alphabet except the forbidden six. As in the previous expression, the space followed by the question mark allows for a space or no space in the middle of the code.

The following program tests Canadian postal codes for an invalid pattern, including the specific rules for each of the letters in the code:

Program 2.3: Checking for Invalid Canadian Postal Codes

```
title "Testing the Regular Expression for Canadian Postal Codes";

data _null_;
   First =  "/[ABCEFGHJ-NPQRSTVXY][0-9]";
   Second = "[ABCEFGHJ-NPRSTV-Z] ?[0-9]";
   Third =  "[ABCEFGHJ-NPRSTV-Z][0-9]/";
   file print;
   input CPC $7.;
   Regex = First||Second||Third;
   if not prxmatch(Regex,CPC) then
      put "Invalid Postal Code " CPC;
datalines;
A1B2C3
ABCDEF
A1B 2C3
12345
D5C6F7
;
```

This program uses the same technique as Program 2.2, with a regular expression designed to identify invalid Canadian postal codes. Because the regular expression is so long (it could be written on a single line, but you would have to scroll right and left to read it), it is broken up into three pieces and these pieces are then concatenated (using the || operator) to create the regular expression.

The output (below) shows the three postal codes that do not conform to the rules:

Figure 2.3: Listing of Invalid Canadian Postal Codes

```
Testing the Regular Expression for Canadian Postal Codes

Invalid Postal Code ABCDEF
Invalid Postal Code 12345
Invalid Postal Code D5C6F7
```

The last postal code in this list was rejected because the letter D is one of the forbidden letters.

Searching for Invalid Email Addresses

It is quite difficult to write an expression for a valid email address. If you use Google to search for a regular expression (for example, "Show me a regex for a valid email address"), you will be overwhelmed with suggestions, some either wrong or inaccurate. This author has found that the most accurate and useful suggestions for regular expressions are on a site called stackoverflow.com.

Some of the suggested regular expressions found online range from something as simple as:

```
/.+@.+/
```

to expressions that contain thousands of characters. The expression above uses the period (a placeholder for any character) and the plus sign repletion operator. The plus sign says to repeat the previous expression one or more times. Therefore, this expression matches just about any string with an @ sign in it.

Some of the very long expressions list all of the valid domains (for example, .com, .org). You can use a simple expression that validates about 99% of email addresses or very complicated expressions that do not allow a single invalid email address to pass through. The following expression is a good compromise between very simple and very complicated:

```
/\b[A-Z0-9._%+-]+@[A-Z0-9.-]+\.[A-Z]{2,}\b/
```

The \b is a word boundary (either the beginning of a string, the end of a string, or a blank space following a word character). The square brackets provide a list of possible characters in that location. The {2,} in the expression means two or more times. You might want to limit the last part of the email address that contains the domain name (such as .com or .edu) with some reasonable number like {2,6}. Also, if you are searching for an email address in a variable, you might want to begin the string with '^' (stands for the beginning of a line).

The program below tests this expression on some valid and invalid email addresses (note that the last repetition operator was written as {2,6}. The 'i' modifier (ignore case) placed after the final delimiter is an instruction to ignore case. Therefore, both upper- and lowercase letters can be included in the email address.

Program 2.4: Testing the Regular Expression for an Email Address

```
*Program to test the Regex for Email Addresses;

title "Testing the Regular Expression Email Addresses";
data _null_;
   file print;
   input Email $50.;
   if not prxmatch("/\b[A-Z0-9._%+-]+@[A-Z0-9.-]+\.[A-Z]{2,6}\b/i",
     Email) then
     put "Invalid Email Address " Email;
datalines;
Jeff.Clark@google.com
no_at_sign_here
1234567890.1234567
fred@rr.tt.org
Bill_Baker@Kerrville.edu
A.B.C@def.too_long
;
```

Here is the output:

Figure 2.4: Output from Program 2.4

```
Testing the Regular Expression Email Addresses

Invalid Email Address no_at_sign_here
Invalid Email Address 1234567890.1234567
Invalid Email Address A.B.C@def.too_long
```

The last email address in the list was rejected because the domain name was too long. If you want to allow longer domain names, leave off the '6' in the last repetition operator or use a larger value.

Verifying Phone Numbers

Your database might store phone numbers in a particular form such as *(nnn)nnn-nnnn*. Another possibility is *nnn.nnn.nnnn*. Writing a regular expression for either of these forms is straightforward (even for this author who is a novice). For the first form, you can use:

```
/\(\d\d\d\)\d\d\d-\d{4}/
```

Both the open and closed parentheses are characters that have a special meaning in a regular expression so that you need to place a backslash (\) before them, as you did with the period in the expression for valid Dx codes in Program 2.1.

To match the second form of a phone number (the one where the digits are separated by periods), you can use:

```
/\d\d\d\.\d\d\d\.\d{4}/
```

As in Program 2.1, you need to place the backslash before each period to indicate that you are referring to a period and not its special meaning.

Converting All Phone Numbers to a Standard Form

Your database might contain phone numbers in many different forms. Table 2.1 is an example:

Table 2.1: Phone Numbers in a Variety of Forms

```
(908)123-1234
609.455-7654
2107829999
(800) 123-4567
```

The following program reads these phone numbers and standardizes them to the form (*nnn*)*nnn-nnnn*:

Program 2.5: Standardizing Phone Numbers

```
*Program to Standardize Phone Numbers;

data Standard_Phone;
   input Phone $16.;
   Digits = compress(Phone,,'kd');
   Phone = cats('(',substr(Digits,1,3),')',substr(Digits,4,3),
      '-',substr(Digits,7));
   drop Digits;
datalines;
(908)123-1234
609.455-7654
2107829999
(800) 123-4567
run;

title "Listing of Standardized Phone Numbers";
proc print data=Standard_Phone noobs;
run;
```

The first step is to use the COMPRESS function to extract all the digits from the phone number. (Remember, 'kd' says to "**k**eep the **d**igits.") Next, the CATS function takes each of the arguments, strips any leading or trailing blanks, and concatenates them. You use the SUBSTR function to extract the area code, the next three digits, and the final four digits and then create phone numbers in the desired form. Here is the output:

Figure 2.5: Listing of Standardized Phone Numbers

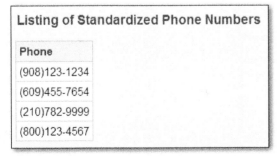

Listing of Standardized Phone Numbers
Phone
(908)123-1234
(609)455-7654
(210)782-9999
(800)123-4567

This author gets especially annoyed when web sites insist that you enter a phone number in a specific form. A few lines of programming can easily standardize any entry to whatever form the web site requires.

Developing a Macro to Test Regular Expressions

Although it's easy to use any of the programs in this chapter that test regular expressions and substitute your own regular expression and string, you might find it more convenient to let a macro do all the work for you.

You can place the macro listed below in your macro library and use it any time you want to test a regular expression. Here is the macro:

Program 2.6: Presenting a Macro to Test Regular Expressions

```
%macro Test_Regex(Regex=, /*Your regular expression*/
                  String= /*The string you want to test*/);
   data _null_;
      file print;
      put "Regular Expression is: &Regex " /
          "String is: &String";
      Position = prxmatch("&Regex","&String");
      if position then put "Match made starting in position " Position;
      else put "No match";
   run;
%Mend Test_Regex;

/*Sample calls
%Test_Regex(Regex=/cat/,String=there is a cat there)
%Test_Regex(Regex=/([A-Za-z]\d){3}\b/, String=a1b2c3)
%Test_Regex(Regex=/([A-Za-z]\d){3}\b/, String=1a2b3c)
*/
```

Testing the macro with the sample calls results in the following three listings:

Figure 2.6: Running the Macro with Test Data

```
Regular Expression is: /cat/
String is: there is a cat there
Match made starting in position 12
```

```
Regular Expression is: /([A-Za-z]\d){3}\b/
String is: a1b2c3
Match made starting in position 1
```

```
Regular Expression is: /([A-Za-z]\d){3}\b/
String is: 1a2b3c
No match
```

You can find this macro and all the programs in this book by going to my author site:

```
support.sas.com/cody
```

Conclusions

The purpose of this chapter is to show that regular expressions have a valid role in testing that data values match a particular pattern. You can either become an expert at using regular expressions, search the web for regular expressions, or, best yet, know a friend or relative who is an expert!

Chapter 3: Standardizing Data

Introduction

A common problem in data processing is a lack of consistency. You might have the name 'Ron Cody' in one file and 'Ronald Cody' or 'Ronald P Cody' in another. You might have a variable such as Gender with some records in uppercase and others in lowercase. These are just a few examples where data cleaning is a necessary first step.

This chapter discusses ways of standardizing names and performing what is called *fuzzy matching*—matching two records based on text values that might not be identical.

Using Formats to Standardize Company Names

Although this example uses company names, any type of data can be standardized using the techniques described in this section.

Suppose you have a SAS data set called Company, containing the names of several different companies. A program to create this data set is shown below:

Program 3.1: Creating the Company Data Set

```
data Company;
   input Name $ 50.;
datalines;
International Business Machines
International Business Macnines, Inc.
IBM
Little and Sons
Little & Sons
Little and Son
MacHenrys
```

```
McHenrys
MacHenries
McHenry's
Harley Davidson
;
```

Here is a listing of data set Company:

Figure 3.1: Listing of Data Set Company

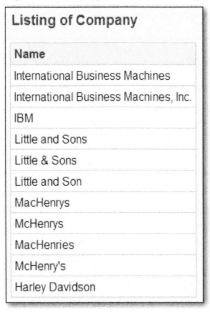

Listing of Company
Name
International Business Machines
International Business Macnines, Inc.
IBM
Little and Sons
Little & Sons
Little and Son
MacHenrys
McHenrys
MacHenries
McHenry's
Harley Davidson

The first step is to decide on a standard name for each company. In this example, let's use 'International Business Machines', 'Little and Sons', and 'McHenrys' as the standard names. There is only one observation with 'Harley Davidson', so we can leave that one alone.

Next, create a format that maps each of the alternative company names in the list to the standard name:

Program 3.2: Creating a Format to Map the Company Names to a Standard

```
proc format;
   value $Company
      "International Business Machines, Inc." =
      "International Business Machines"
      "IBM" = "International Business Machines"
      "Little & Sons" = "Little and Sons"
      "Little and Son" = "Little and Sons"
      "MacHenrys"      = "McHenrys"
      "MacHenries"     = "McHenrys"
      "McHenry's"      = "McHenrys";
run;
```

You can use a PUT function to convert the alternative names to standard names as follows:

Program 3.3: Using a PUT Function to Standardize the Company Names

```
data Standard;
   set Company;
   Standard_Name = put(Name,$Company.);
run;

title "Listing of Standard";
proc print data=Standard noobs;
run;
```

The PUT function, described in Chapter 1, takes the value of the first argument (Name in this example) and formats it using the format specified in the second argument ($Company in this example). The result is that the company names are all mapped to the standard names. See the listing below:

Figure 3.2: Listing from Program 3.3

Listing of Standard

Name	Standard_Name
International Business Machines	International Business Machines
International Business Macnines, Inc.	International Business Macnines
IBM	International Business Machines
Little and Sons	Little and Sons
Little & Sons	Little and Sons
Little and Son	Little and Sons
MacHenrys	McHenrys
McHenrys	McHenrys
MacHenries	McHenrys
McHenry's	McHenrys
Harley Davidson	Harley Davidson

Inspection of this listing shows that all the company names have been standardized.

Creating a Format from a SAS Data Set

If you already have the standard company names and a set of alternative names in a SAS data set, you can use that data set to create the $Company format using what PROC FORMAT calls a *control data set* (keyword: CNTLIN=). To demonstrate this, suppose you have a data set with the standard and alternative names. Program 3.4 (below) creates a data set called Standardize that contains a standard name and several alternative names:

Program 3.4: Creating Data Set Standardize

```
*Create Data Set Standardize;
data Standardize;
   input @1  Alternate $40.
         @41 Standard  $40.;
datalines;
International Business Machines, Inc.  International Business Machines
IBM                                    International Business Machines
Little & Sons                          Little and Sons
Little and Son                         Little and Sons
MacHenrys                              McHenrys
MacHenries                             McHenrys
McHenry's                              McHenrys
;
```

Here is a listing of data set Standardize:

Figure 3.3: Listing of Data Set Standardize

Listing of Data Set Standardize

Alternate	Standard
International Business Machines, Inc.	International Business Machines
IBM	International Business Machines
Little & Sons	Little and Sons
Little and Son	Little and Sons
MacHenrys	McHenrys
MacHenries	McHenrys
McHenry's	McHenrys

Control data sets have special requirements for the variable names that hold the value to be formatted, the label associated with this value, the format name, and whether the format is character or numeric. The variable names required for a control data set are:

Start
> The starting value in a range. If there is only one value (as in this case), you do not have to include a value for End in the control data set.

End
> An ending value if you have a range such as 10 to 20 (10 would be the start value and 20 would be the end value).

Label
> The format label.

Fmtname
> The name of the format you want to create. Do not include a dollar sign in the name, even if this is a character format.

Type
> The type of the format you want to create. Use a 'C' if you are creating a character format; use an 'N' if you are creating a numeric format.

To create a format called $Company with the variable Alternate as the value to be formatted and Standard as the name to act as the format label, run the following DATA step:

Program 3.5: DATA Step to Convert the Standardize Data Set Into the Control Data Set

```
data Control;
    set Standardize(rename=(Alternate=Start Standard=Label));
    retain Fmtname "Company" Type "C";
run;
```

Because the variable Alternate is the value you want to format, you rename this variable to Start. The variable Standard is renamed to Label. The values of Fmtname and Type are set using a RETAIN statement. You could have used assignment statements for this, but the RETAIN statement executes only once and is therefore more efficient (and elegant).

Here is a listing of data set Control:

Figure 3.4: Listing of Data Set Control

Listing of Data Set Control

Start	Label	Fmtname	Type
International Business Machines, Inc.	International Business Machines	Company	C
IBM	International Business Machines	Company	C
Little & Sons	Little and Sons	Company	C
Little and Son	Little and Sons	Company	C
MacHenrys	McHenrys	Company	C
MacHenries	McHenrys	Company	C
McHenry's	McHenrys	Company	C

You are ready to run PROC FORMAT, using the Control data set to create your format. Here are the PROC FORMAT statements to create the $Company format:

Program 3.6: Using PROC FORMAT with the Control Data Set

```
proc format library=work cntlin=Control fmtlib;
run;
```

You name the control data set following the PROC FORMAT option CNTLIN= and, optionally, add the option FMTLIB. This option is an instruction to list all the formats in the library you named on the LIBRARY= option. You would probably be placing this format in a permanent library (described in Chapter 1) and not the Work library. Including the FMTLIB statement produces the following format description:

Figure 3.5: Listing as a Result of Including the FMTLIB Option

```
Output from the FMTLIB Option


         FORMAT NAME: $COMPANY LENGTH:   31   NUMBER OF VALUES:    7
    MIN LENGTH:   1  MAX LENGTH:  40  DEFAULT LENGTH:  31  FUZZ:        0

  START              END               LABEL    (VER. V7|V8   09NOV2016:16:02:49)

  IBM                IBM               International Business Machines
  International Bu   International Bu   International Business Machines
  Little & Sons      Little & Sons     Little and Sons
  Little and Son     Little and Son    Little and Sons
  MacHenries         MacHenries        McHenrys
  MacHenrys          MacHenrys         McHenrys
  McHenry's          McHenry's         McHenrys
```

Some of the names are truncated in this listing, but this format is equivalent to the one you created in Program 3.2.

Using TRANWRD and Other Functions to Standardize Addresses

TRANWRD (stands for translate word) is useful in data cleaning operations. The syntax for this function is as follows:

New_String = TRANWRD (*String*, *Find*, *Replace*) ;

where *String* is the value you want to search, *Find* is the string you are searching for, *Replace* is the replacement string, and *New_String* is the result. If the length of *New_String* has not been previously set, a default length of 200 will be used.

For efficiency, you should explicitly set the length of *New_String* because if the length of *New_String* defaults to 200, this could significantly increase the amount of space needed to store the resulting data set.

For example, if String = "12 River Road", then if you use TRANWRD like this:

```
    String = TRANWRD(String,"Road","Rd.");
```

String will have the value "12 River Rd.".

A small SAS data set called Addresses was created to demonstrate how the TRANWRD function can help standardize an address. Here is a listing of data set Addresses:

In case you want to play along, run Program 3.7 to create the Clean.Addresses data set:

Program 3.7: Creating the Clean.Addresses Data Set

```
*Program to create data set Addresses;
data Clean.Addresses;
   input #1 Name      $25.
         #2 Street    $30.
```

```
            #3 @1   City   $20.
               @21 State  $2.
               @23 Zip    $10.;
datalines;
Robert L. Stevenson
12 River Road
Hartford            CN06101
Mr. Fred Silver
145A Union Court
Flemingron          NJ08822
Mrs. Beverly Bliss
6767 Camp Verde Road East
Center   Point       tx78010
Mr. Dennis Brown, Jr.
67 First Street
Miami               FL33101
Ms. Sylvia D'AMORE
23 Upper Valley Rd.
Clear    Lake        WI54005
;
```

Here is the listing:

Figure 3.6: Listing of the Addresses Data Set

```
Listing of Data Set Addresses

       Name              Street                    City          State   Zip

Robert L. Stevenson    12 River Road             Hartford        CN     06101
Mr.  Fred Silver       145A Union Court          Flemingron      NJ     08822
Mrs. Beverly Bliss     6767 Camp Verde Road East Center  Point   tx     78010
Mr. Dennis Brown, Jr.  67 First Street           Miami           FL     33101
Ms. Sylvia D'AMORE     23 Upper Valley Rd.       Clear   Lake    WI     54005
```

Some of the values in this file contain multiple spaces where you only expect a single space (for example, the last City value). Also, in the third observation, the state abbreviation is in lowercase and in the last observation, the last name is in uppercase. You need to correct these problems and convert the names 'Road', 'Court', and 'Street' to standard abbreviations. You might also want to remove the comma and 'Jr.' from the fourth address. The program below accomplishes all these tasks:

Program 3.8: Beginning Address Standardization

```
data Std_Addresses;
   set Clean.Addresses;
   array Chars[*] Name Street City;  ❶

   do i = 1 to dim(Chars);  ❷
      Chars[i] = compbl(Chars[i]);  ❸
      Chars[i] = propcase(Chars[i]," '");  ❹
   end;

   Street = tranwrd(Street,"Road","Rd.");  ❺
   Street = tranwrd(Street,"Court","Ct.");
   Street = tranwrd(Street,"Street","St.");
```

```
      State = Upcase(State);
      *Remove ,Jr. from Name;
      Name = tranwrd(Name,"Jr.","  ");  ❻
      Name = tranwrd(Name,",","  ");  ❼
      drop i;
run;


title "Listing of Std_Addresses";
proc print data=Std_Addresses noobs;
run;
```

❶ Make an array of the three variables on which you plan to apply the COMPBL and PROPCASE functions.

❷ Although it's not too hard to count to three, in a setting where there are many more variables, using the DIM function (which returns the number of elements in the array) makes counting unnecessary.

❸ Use the COMPBL function to replace multiple blanks with a single blank.

❹ Use the PROPCASE function to convert all three variables to proper case. Using a blank and single quote mark as delimiters ensures that the name D'AMORE is changed to D'Amore.

❺ Use the TRANWRD function to convert 'Road', 'Court', and 'Street' to the appropriate abbreviations. In a more realistic situation, you would include many more names such as Boulevard or Lane. Later in this chapter, you will see a more compact method, using the PRXCHANGE function, to address this problem (pun intended).

❻ and ❼ You can use the TRANWRD function to change 'Jr.' and the comma to a blank.

Below is the listing of the standardized data set:

Figure 3.7: Listing of Data Set Std_Addresses

Listing of Std_Addresses

Name	Street	City	State	Zip
Robert L. Stevenson	12 River Rd.	Hartford	CN	06101
Mr. Fred Silver	145a Union Ct.	Flemingron	NJ	08822
Mrs. Beverly Bliss	6767 Camp Verde Rd. East	Center Point	TX	78010
Mr. Dennis Brown	67 First St.	Miami	FL	33101
Ms. Sylvia D'Amore	23 Upper Valley Rd.	Clear Lake	WI	54005

Using Regular Expressions to Help Standardize Addresses

An alternative to changing names like 'Street' and 'Road' to their respective abbreviations is to remove all such words from the address. You can use the TRANWRD function to do this, but you would need a separate statement for each substitution. A more compact way to remove all references to streets and roads is to use the PXCHANGE function. Program 3.9, shown next, illustrates how this works:

Program 3.9: Using PRXCHANGE to Replace Street and Road with Blanks

```
data Remove_Names;
   set Clean.Addresses(keep=Street);
   Original = Street;
   Words = "s/\sRoad\b|\sCourt\b|\sStreet\b";
   Abbreviations = "|\sRd\.\s*$|\sCt\.\s*$|\sSt\.\s*$/ /";
   Regex = cats(Words,Abbreviations);
   Street = prxchange(Regex,-1,Street);
   Keep Original Street;
run;
```

The variable Original is set to the value of 'Street' before the replacements, so that you can see the values before and after the PRXCHANGE function does its magic. Because the regular expression is a bit long, you break it into two pieces (variables Words and Abbreviations) and use the CATS (concatenation) function to create the final regular expression (Regex). Next, you use the PRXCHANGE function to substitute one string for another.

The regular expression that searches for the words and abbreviations is a bit complicated. You can take it on faith that it works or read the explanation that follows:

The 's' before the first delimiter is needed when you are making substitutions. Each term in the regular expression begins with \s (a white space character—blanks, tabs, and several other white space characters) and ends with either a \b (a word boundary) or a \s*$. The asterisk (*) is a repetition operator that repeats the previous expression 0 or more times and the dollar sign ($) means end-of-line. Therefore, the expression \s*$ represents 0 or more white space characters followed by the end of the line.

Because a period has a special meaning in regular expressions (stands for any character), you need to precede it with a backspace, informing the function that you are treating the period as a character and not a placeholder. Following each of the abbreviations (Rd., Ct., and St.) are the characters \s*$. The reason you can't use \b following any of the abbreviations is that \b is a word boundary and needs to follow a word character. However, a period is not a word character so you need to use an alternative expression such as \s*$.

The vertical bar (referred to as a pipe symbol) between each term in the expression says to match any one of the expressions. You place a blank space following the regular expression so that any of the terms in the list will be changed to a blank (this blank is included in the variable Regex). Use a -1 as the next argument to tell the function to look to the end of the string for matches. The last argument in the function provides the name of the variable where the substitutions are to take place.

Running Program 3.9 results in the following output:

Figure 3.8: Output from Program 3.9

Road, Court, Street, Rd., etc. Converted to Blanks	
Original	Street
12 River Road	12 River
145A Union Court	145A Union
6767 Camp Verde Road East	6767 Camp Verde East
67 First Street	67 First
23 Upper Valley Rd.	23 Upper Valley

The words 'Road', 'Court', and 'Street' have been removed from the variable Street.

Performing a "Fuzzy" Match between Two Files

Because of confidentiality laws or other reasons, you might find yourself needing to merge two files where you do not have a unique identifier. You might need to merge the files based on several variables, including people's names. This causes problems because names do not always provide an exact match. One file might include a middle initial and the other not. There mgiht be spelling errors in one or both files. The SPEDIS function is quite useful in a situation like this.

The SPEDIS function (stands for spelling distance) measures the spelling distance between two strings. The syntax is:

```
Spelling_Distance = spedis(String1, String2);
```

where *String1* and *String2* are the strings you want to compare.

If the strings are exactly the same, the function returns a 0. If the strings are not the same, the function uses an algorithm that give penalty points to various types of spelling errors. The most penalty points are assigned when the first letters do not match in the two strings. Other spelling errors (such as interchanging two letters, adding a letter, or missing a letter) are given other values. The function adds up the total penalty points, but it is not finished yet. The final step is to compute the total penalty points as a percentage, based on the length of the first string. This makes sense: A single letter mistake in a three-letter word is a pretty severe error—a single letter mistake in a 10-letter word is pretty good! You can investigate observations where variables such as Gender or DOB match and whose names are within a predetermined spelling distance.

To demonstrate this technique, you have two SAS data sets, Discharge (data on patients discharged from a hospital) and MI (patients who had a myocardial infarction—heart attack). Both contain the patient's name, date of birth, and gender. You want to determine if some of the discharged patients are in the heart attack data set. Listings of both data sets are shown next:

Figure 3.9: Data Set Discharge

Listing of Data Set Discharge

LastName	DOB	Gender
Smith	10/21/1955	M
DANIELS	11/11/1944	F
O'Brien	12/25/1951	M
Chen	06/05/1965	F
Stevenson	01/07/1970	F

Figure 3.10: Data Set MI

Listing of Data Set MI

LastName_MI	DOB_MI	Gender_MI
Smyth	10/21/1955	M
Smith	10/20/1945	F
Daniels	11/11/1944	F
Chien	06/05/1965	F
swanson	01/07/1970	M

Your goal is to find patients in both data sets who have the same gender and date of birth and where the spelling distance between the two names is less than a predetermined value. Before you search for matches, it is a good idea to apply one of the case functions such as PROPCASE to ensure that you have consistent cases in both files. If any of the character variables contain blanks, you might also want to apply the COMPBL (compress blanks) function as well. The program listed next ensures that names are in proper case and gender is in uppercase:

Program 3.10: Using the PROPCASE and UPCASE Functions to Ensure Consistent Cases

```
data Clean.Discharge;
   set Clean.Discharge;
   LastName = propcase(LastName," '");
   Gender = Upcase(Gender);
run;

data Clean.MI;
   set Clean.MI;
   LastName_MI = propcase(LastName_MI," '");
   Gender_MI = upcase(Gender_MI);
run;
```

You can use PROC SQL to create a Cartesian product—a data set that contains every combination of variables from the two data sets. A program to produce a Cartesian product using the two data sets Discharge and MI is shown in Program 3.11:

Program 3.11: Creating a Cartesian Product

```
title "Creating a Cartesian Product";
proc sql;
   create table Clean.join as
   select *
   from Clean.Discharge, Clean.MI;
   /* A WHERE Clause will go here -
      Do NOT run this program without it
   */
quit;

title "Listing of Data Set JOIN";
proc print data=Clean.join;
   id LastName;
run;
```

Here is a partial listing of the resulting Cartesian product (without a WHERE clause):

Figure 3.11: Partial Listing of Data Set Join

Listing of Data Set JOIN

LastName	DOB	Gender	LastName_MI	DOB_MI	Gender_MI
Smith	10/21/1955	M	Smyth	10/21/1955	M
Smith	10/21/1955	M	Smith	10/20/1945	F
Smith	10/21/1955	M	Daniels	11/11/1944	F
Smith	10/21/1955	M	Chien	06/05/1965	F
Smith	10/21/1955	M	Swanson	01/07/1970	M
Daniels	11/11/1944	F	Smyth	10/21/1955	M
Daniels	11/11/1944	F	Smith	10/20/1945	F
Daniels	11/11/1944	F	Daniels	11/11/1944	F
Daniels	11/11/1944	F	Chien	06/05/1965	F
Daniels	11/11/1944	F	Swanson	01/07/1970	M
O'Brien	12/25/1951	M	Smyth	10/21/1955	M
O'Brien	12/25/1951	M	Smith	10/20/1945	F
O'Brien	12/25/1951	M	Daniels	11/11/1944	F
O'Brien	12/25/1951	M	Chien	06/05/1965	F
O'Brien	12/25/1951	M	Swanson	01/07/1970	M
Chen	06/05/1965	F		10/21/1955	M

The size of the Cartesian product is the number of observations in the first data set times the number of observations in the second data set. Even if both data sets are relatively small, say 10,000 observations each, the result would be very large (100,000,000). Because the two data sets used in this example are so small, it was possible to produce and list the results.

The next step is to use PROC SQL to search for exact matches and matches where the names are similar. First, here is the program to list exact matches. (Note: This program ignores missing values—a more robust program would take missing values into account.)

Program 3.12: Using PROC SQL to Search for Exact Matches

```
proc sql;
   create table Clean.Exact as
   select *
   from Clean.Discharge, Clean.MI
   where DOB eq DOB_MI          and
         Gender eq Gender_MI    and
         LastName = LastName_MI;
quit;
```

Next, the program to search for patients who have the same gender and date of birth and whose names are similar:

Program 3.13: Using PROC SQL to Find Matches with Similar Names

```
proc sql;
   create table Clean.Possible as
   select *
   from Clean.discharge, Clean.MI
   where DOB eq DOB_MI          and
         Gender eq Gender_MI    and
         0 lt spedis(LastName,LastName_MI) le 25;
quit;
```

A spelling distance of 25 was used for this example. If you make this value too small, you might miss possible matches—it you make this value too large, you might match subjects who should not be matched. You will need to test different values to see if you are over or under matching. The two data sets, Possible and Exact, are listed below:

Figure 3.12: Listing of Data Sets Possible and Exact

Listing Exact Matches

LastName	DOB	Gender	LastName_MI	DOB_MI	Gender_MI
Daniels	11/11/1944	F	Daniels	11/11/1944	F

Listing of Possible Matches

LastName	DOB	Gender	LastName_MI	DOB_MI	Gender_MI
Smith	10/21/1955	M	Smyth	10/21/1955	M
Chen	06/05/1965	F	Chien	06/05/1965	F

Conclusions

This chapter just scratched the surface in describing some of the steps to standardize data. If you search the web for terms such as "standardizing addresses AND SAS," you will find a wealth of information on this topic.

Chapter 4: Data Cleaning Techniques for Numeric Data

Introduction

Numeric data presents a very different challenge from character data. This chapter covers several techniques for identifying possible errors in numeric data. One of the first steps with any numeric variable is to look at the highest and lowest values, both in table form and with the aid of histograms or other graphical techniques. You can specify reasonable ranges for some categories of numeric data, such as physiological variables. For example, finding a value of 900 for a heart rate would most certainly indicate a data error. This chapter will introduce programs and macros for numeric data for which range checking is feasible. For other categories of numeric data, you can use statistical techniques to determine possible outliers, the topic of the next chapter. Let's get started.

Using PROC UNIVARIATE to Examine Numeric Variables

Before you run any statistical analysis on numeric variables, you should first use PROC UNIVARIATE to create both tabular and graphical information on these variables. The SAS data set called Patients, described in Chapter 1, will be used to demonstrate the data cleaning techniques in this chapter. To help refresh your memory, the figure shown next lists the first 10 observations in this data set:

Figure 4.1: Listing the First 10 Observations in the Patients Data Set

Listing of the First 10 Observations in Data Set Patients

Patno	Account_No	Gender	Visit	HR	SBP	DBP	Dx	AE
	DE56405	M	06/15/2010	87	128	98	195.920	0
001	CT14882	M	06/12/2012	69	124	86	713.410	0
002	MD78461	M	06/04/2010	76	130	80	047.570	1
003	DE51381	f	06/22/2013	70	56	70	108.510	0
004	CT37146	M	05/18/2013	76	112	84	669.860	0
005	DE00080	F	04/08/2012	91	106	84	078.160	0
005	DE00080	F	04/08/2012	91	106	84	078.160	0
006	DE37709	M	07/27/2014	71	104	88	967.570	0
007	VT56383	F	01/13/2014	63	128	80	640.260	1
007	NJ90043	M	08/06/2010	83	130	102	564.870	0

Let's use the numeric variables HR (heart rate), SBP (systolic blood pressure), and DBP (diastolic blood pressure) to demonstrate data cleaning techniques for numeric variables. As mentioned in the introduction, a good first step is to run PROC UNIVARIATE.

Program 4.1: Running PROC UNIVARIATE on HR, SBP, and DBP

```
title "Running PROC UNIVARIATE on HR, SBP, and DBP";
proc univariate data=Clean.Patients;
   id Patno;
   var HR SBP DBP;
   histogram / normal;
run;
```

It is important to include an ID statement when you run this procedure. The most important reason is to identify the ID of any patients who have suspicious values. The variable Patno (patient number) is the identifying variable in the Patients data set. You use a VAR statement to list the variables you want to examine. If you omit the VAR statement, PROC UNIVARIATE will analyze every numeric variable in your data set, usually not a desirable approach. The HISTOGRAM statement is a request to create a histogram for all the variables listed on the VAR list. If you want to superimpose a normal distribution on the histogram, include the NORMAL option on the HISTOGRAM statement. You can include a list of variables on the HISTOGRAM statement if you only want histograms for selected variables. Because PROC UNIVARIATE produces many pages of output, the output displayed in this chapter is divided into sections and only output for SBP is displayed (to save space):

Figure 4.2: PROC UNIVARIATE Output (for SBP only)

Running PROC UNIVARIATE on HR, SBP, and DBP

Variable: SBP (Systolic Blood Pressure)

Moments			
N	101	Sum Weights	101
Mean	121.881188	Sum Observations	12310
Std Deviation	24.4177342	Variance	596.225743
Skewness	4.22443829	Kurtosis	29.575593
Uncorrected SS	1559980	Corrected SS	59622.5743
Coeff Variation	20.0340467	Std Error Mean	2.42965536

The first part of the output contains a number of statistical values that are not particularly useful at this point. You will want to revisit this output once you have ensured that you have removed or corrected data errors. At this time, the only value you may be interested in is N, the number of non-missing observations.

PROC UNIVARIATE Output (continued)

Basic Statistical Measures			
Location		Variability	
Mean	121.8812	Std Deviation	24.41773
Median	120.0000	Variance	596.22574
Mode	130.0000	Range	244.00000
		Interquartile Range	18.00000

As with the previous snippet of output, many of these values are not meaningful until the data cleaning operation has been completed. Values in this portion of the output include measures of central tendency (mean, median, and mode) as well as measures of variability (standard deviation, variance, range, and interquartile range).

PROC UNIVARIATE Output (continued)

Tests for Location: Mu0=0				
Test		Statistic	p Value	
Student's t	t	50.16398	Pr > |t|	<.0001
Sign	M	50.5	Pr >= |M|	<.0001
Signed Rank	S	2575.5	Pr >= |S|	<.0001

The statistical tests shown in this section of output are testing the null hypothesis that the population mean from which the sample was taken is equal to 0. You can test a different null hypothesis by including a PROC UNIVARIATE option called MU0=value, where value is a population mean you are interested in.

PROC UNIVARIATE Output (continued)

Quantiles (Definition 5)	
Level	Quantile
100% Max	300
99%	210
95%	148
90%	132
75% Q3	128
50% Median	120
25% Q1	110
10%	104
5%	100
1%	92
0% Min	56

This portion of the output lists the value of your variable (SBP in this example) for several quantile values. The 0% and 100% values represent the lowest and highest values of SBP. For a healthy adult, neither of these values is reasonable and they probably represent data errors. Other values of interest are the 25th, 50th, and 75th percentiles. The value at the 50th percentile is also called the *median*. Half of the SBP values are greater than this value and half of the SBP values are less than this value. It is quite possible that the value of the median in this listing will not change after you have eliminated the data errors.

PROC UNIVARIATE Output (continued)

Extreme Observations					
Lowest			Highest		
Value	Patno	Obs	Value	Patno	Obs
56	003	4	148	060	60
92	016	17	152	066	66
94	038	37	160	013	15
98	089	89	210	019	20
98	074	74	300	023	23

Finally, a section of output that represents a useful data cleaning tool: the five lowest and five highest values of SBP. Notice the column labeled Patno. It identifies the patient number for each observation in the table. Without an ID statement, only the Obs (observation number) column would be shown, along with the value of SBP. An inspection of this table highlights some suspicious values. Certainly the values of 56 for patient 003 and 300 for patient 023 need to be checked. You might also want to double-check the value of 210 for patient 019.

PROC UNIVARIATE Output (continued)

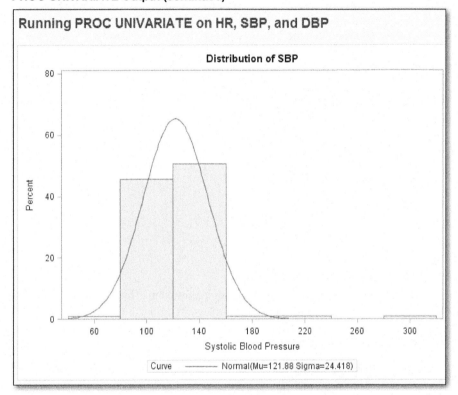

The histogram in this last section of output shows very clearly that there are possible errors at the low and high ends of the distribution.

Describing an ODS Option to List Selected Portions of the Output

In the data cleaning phase, much of the output from PROC UNIVARIATE is not useful. You can use an ODS SELECT statement to specify which portion or portions of the output you want listed. In order to understand how this works, here is a brief discussion of the output delivery system.

Prior to SAS 7, all SAS output was managed separately by each procedure. That was fine if all you needed was a SAS produced listing. However, about that time, other forms of output such as HTML, PDF, and RTF (rich text format) were becoming popular. In order to accommodate all these different output destinations, SAS made a major change to its software by introducing the output delivery system. Instead of having each procedure create and format its output, each procedure produced output objects. Basically, these output objects were data sets that contained all of the information generated by each procedure. The output delivery system was designed to take each of the output objects and send output to the destinations listed earlier.

Each procedure produces one or more output objects. If you know the name of the output object that contains the portion of the output you want, you can use the following statement:

```
ODS SELECT output-object-name;
```

where *output-object-name* is the name of one or more output objects that you want to display. The next obvious question is how do you know the names of the output objects? There are several ways: If you are working in SAS Display Manager, you can open the Results window to see the following:

Figure 4.3: Open the Results Window

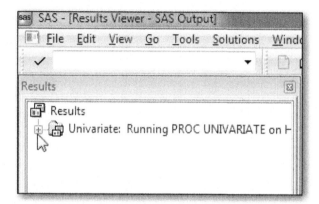

Next, click on the plus (+) sign to the left of the procedure you are interested in. This expands to show you the three variables that were analyzed.

Figure 4.4: Further Expanding the Results Window

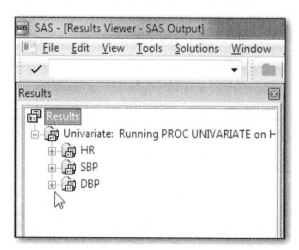

Choose any variable and further expand the Results window to see a list of output objects (Figure 4.5):

Figure 4.5: List of Output Objects Produced by PROC UNIVARIATE

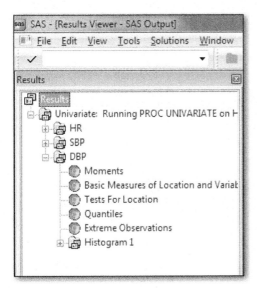

This list shows you the descriptions of the output objects produced by PROC UNIVARIATE. Click on the name of any output object and the output window will jump to that portion of the output. Once you determine which output object you want, right-click on it to see the name of the object (Figure 4.6):

Figure 4.6: Properties of the Output Object Showing Name and Description

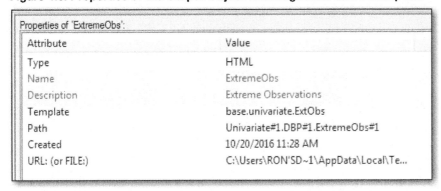

The name of the output object associated with the description "Extreme Observations" is ExtremeObs. If you only want to see the extreme observations portion of the output from PROC UNIVARIATE, insert the following statement immediately before the procedure:

```
ODS Select ExtremeObs;
```

When you run the procedure, only the portion containing the extreme observations will be shown.

Listing Output Objects Using the Statement TRACE ON

If you are working in a SAS environment where you do not have a Results window, you can add the statement TRACE ON before the procedure you are interested in and TRACE OFF after the procedure. This will generate a list of output object descriptions and names in the SAS Log. A modification of this statement, TRACE ON/Listing, will place the names and descriptions of the output object in the output, right before the portion of output produced by that output object.

Using a PROC UNIVARIATE Option to List More Extreme Values

If you have a large data set and you expect a lot of errors, you might elect to see more than the five lowest and highest observations. To change the number of extreme observations in the output, add the following procedure option:

NEXTROBS=*number*

where *number* is the number of extreme observations to include in the list. For example, to see the 10 lowest and highest values for your variables, submit the following program:

Program 4.2: Adding the NEXTROBS= Option to PROC UNIVARIATE

```
title "Running PROC UNIVARIATE on HR, SBP, and DBP";
ods select ExtremeObs;
proc univariate data=Clean.Patients nextrobs=10;
    id Patno;
    var HR SBP DBP;
    histogram / normal;
run;
```

This program includes both the ODS SELECT statement as well as the NEXTROBS= option. The output is shown below:

Figure 4.7: Adding the NEXTROBS= Option to PROC UNIVARIATE

Running PROC UNIVARIATE on HR, SBP, and DBP
Adding the Option NEXTROBS=

Variable: SBP (Systolic Blood Pressure)

Extreme Observations					
Lowest			Highest		
Value	Patno	Obs	Value	Patno	Obs
56	003	4	138	059	59
92	016	17	140	062	62
94	038	37	140	065	65
98	089	89	140	073	73
98	074	74	148	029	29
100	083	83	148	060	60
102	061	61	152	066	66
104	100	100	160	013	15
104	096	96	210	019	20
104	088	88	300	023	23

As before, only the output for SBP is displayed here.

Presenting a Program to List the 10 Highest and Lowest Values

The previous section described how to use PROC UNIVARIATE with the NEXTROBS= option to list the top and bottom *n* values of a variable. This section presents an alternative—a program that performs the same task. The purpose of the program is to demonstrate some SAS programming techniques as well as to set the groundwork for developing a macro to print the *n* highest and lowest values of a variable.

It seems pretty obvious how to determine the 10 lowest values of a variable: Just sort the data set by that variable, eliminate any missing values, and print the first 10 observations. Every time I teach my data cleaning workshop (based on the data cleaning book), I ask, "How would you list the 10 highest values?" I have never taught a class where a student did not suggest another PROC SORT with the DSCENDING option so that the data set would be sorted from high to low. While that does work, it is not necessary to re-sort the data set. If you know how many observations of non-missing values there are, you can subtract nine from this value and then list the observations from that point to the end.

There are several ways to determine how many observations are in a SAS data set. The method presented here is the classic method and probably the most straightforward one—using the SET option NOBS=. The program presented below will list the 10 lowest and highest values of HR in the Patients data set:

Program 4.3: Listing the 10 Highest and Lowest Values of a Variable

```
proc sort data=Clean.Patients(keep=Patno HR ❶
                              where=(HR is not missing))
                              out=Tmp;
   by HR;
run;

data _null_;
   if 0 then set Tmp nobs=Number_of_Obs; ❷
   High = Number_of_Obs - 9;
   call symputx('High_Cutoff',High); ❸
   stop; ❹
run;

title "Ten Highest and Lowest Values for HR";
data _null_;
   set Tmp(obs=10)                     /* 10 lowest values   */ ❺
       Tmp(firstobs=&High_Cutoff); /* 10 highest values */
   file print; ❻
   if _n_ le 10 then do; ❼
      if _n_ = 1 then put / "Ten Lowest Values"; ❽
      put "Patno = " Patno @15 "Value = " HR;
   end;
   else if _n_ ge 11 then do; ❾
      if _n_ = 11 then put / "10 Highest Values";
      put "Patno = " Patno @15 "Value = " HR;
   end;
run;
```

❶ Use PROC SORT to sort the Patients data set by HR. Use a KEEP= data set option to bring the two variables Patno and HR into the PDV (program data vector). In addition, you can remove the missing values with a WHERE= data set option. Finally, use the OUT= data set option to send the sorted data to a temporary data set called Tmp.

❷ This is a rather strange statement. Zero is always false, so you might wonder why would you write an IF statement that is never true? It turns out that the data set option NOBS=*variable* executes at **compile** time, where IF statements are not evaluated. The result of this DATA _NULL_ step is to assign the number of observations in data set Patients to a variable called Number_of_Obs.

❸ To list the 10 highest values in the sorted data set, you need to subtract nine from the number of observations. The CALL SYMPUTX CALL routine takes two arguments. The first argument is the name of a macro variable—the second argument is a DATA step variable. The CALL routine assigns the value of the data set variable to the macro variable. After this CALL routine executes, the macro variable High_Cutoff holds the value of High.

❹ You need a STOP statement because you are not reading any observations from the Tmp data set and you will not reach an end-of-file, a process that usually causes a DATA step to stop. Without the STOP statement, this DATA _NULL_ step will loop until it times out.

❺ The trick here is to set the same data set twice—once with the option OBS=10 (stop when you read observation number 10) and once with the option FIRSTOBS=&High_Cutoff (giving you the highest 10 values).

❻ The FILE PRINT statement sends the results of the subsequent PUT statement to the output device (instead of the default location, the SAS Log).

❼ Use the internal variable _N_ to list the first 10 observations.

❽ When you are processing the first observation, print out the message "Ten Lowest Values."

❾ Even though you are jumping to the position specified by the value of &High_Cutoff, the internal counter _N_ is equal to 11.

Listed below is the output from Program 4.3:

Figure 4.8: Output from Program 4.3

```
Ten Highest and Lowest Values for HR


Ten Lowest Values
Patno = 050    Value = 32
Patno = 050    Value = 43
Patno = 061    Value = 47
Patno = 058    Value = 49
Patno = 013    Value = 50
Patno = 041    Value = 50
Patno = 052    Value = 50
Patno = 066    Value = 50
Patno = 036    Value = 52
Patno = 038    Value = 53
```

```
10 Highest Values
Patno = 090    Value = 84
Patno = 099    Value = 85
Patno =        Value = 87
Patno = 080    Value = 88
Patno = 005    Value = 91
Patno = 005    Value = 91
Patno = 044    Value = 92
Patno = 077    Value = 95
Patno = 034    Value = 115
Patno = 045    Value = 900
```

Presenting a Macro to List the *n* Highest and Lowest Values

This section uses the logic of Program 4.3 to produce a macro that lists the *n* highest and lowest (non-missing) values in a data set. First the macro, then the explanation:

Program 4.4: Converting Program 4.3 to a Macro

```
*Macro Name: HighLow
Purpose: To list the "n" highest and lowest values
Arguments: Dsn      - Data set name (one- or two-level)
           Var      - Variable to list
           IDvar    - ID variable
           N        - Number of values to list (default = 10)
```

```
example: %HighLow(Dsn=Clean.Patients,
                  Var=HR,
                  IDvar=Patno,
                  N=7)
;
%macro HighLow(Dsn=,      /* Data set name            */
               Var=,      /* Variable to list         */
               IDvar=,    /* ID Variable              */
               N=10       /* Number of high and low
                             values to list.
                             The default number is 10 */);

   proc sort data=&Dsn(keep=&IDvar &Var
                       where=(&Var is not missing))
                       out=Tmp;
      by &Var;
   run;

   data _null_;
      if 0 then set Tmp nobs=Number_of_Obs;
      High = Number_of_Obs - %eval(&N - 1);
      call symputx('High_Cutoff',High);
      stop;
   run;

   title "&N Highest and Lowest Values for &Var";
   data _null_;
   set Tmp(obs=&N)                    /* 'n' lowest values  */
       Tmp(firstobs=&High_Cutoff); /* 'n' highest values */
   file print;
   if _n_ le &N then do;
      if _n_ = 1 then put / "&N Lowest Values";
      put "Patno = " &IDvar @15 "Value = " &Var;
   end;
   else if _n_ ge %eval(&N + 1) then do;
      if _n_ = %eval(&N + 1) then put / "&N Highest Values";
      put "&IDvar = " &IDvar @15 "Value = " &Var;
   end;
   run;

   proc datasets library=work nolist;
      delete Tmp;
   run;
   quit;
%mend HighLow;
```

For those readers who might be unfamiliar with SAS macros or who have written a few macros but still feel a bit shaky in macro-land (like this author), here is a brief explanation of SAS macros. (If you are a macro maven, just skip the explanation below):

A *macro* is a piece of SAS code where parts of the code are substituted with variable information by the macro processor before the code is processed in the usual way by the SAS compiler. The macro presented here is called HighLow, and it begins with a %MACRO statement and ends with a %MEND (macro end) statement. The first line of the macro contains the macro name, followed by a list of arguments. This style of listing the arguments is called *named parameters*. You list the name of the parameter, followed by an equal sign and, if you wish, a default value. An alternative way to list calling arguments when defining a macro is with *positional parameters* (where it is the order that is important).

When the macro is called (executed), the macro processor replaces every macro variable (variable names preceded by an ampersand) with the values you specify. If you want to use this macro to look for the top and bottom seven values of HR in the Patients data set, call the macro like this:

```
%HighLow(Dsn=Clean.Patients,
         Var=HR,
         Idvar=Patno,
          N=7)
```

The macro processor will substitute these calling arguments for the macro variables in the program. For example, the SET statement will become:

```
set Clean.Patients(keep=Patno HR);
```

&Dsn was replaced by Clean.Patients, &IDvar was replaced by Patno, and &Var was replaced by HR. To be sure this concept is clear (and to help you understand how the macro processor works), you can call the macro with the MPRINT option turned on. This option displays the macro code in the SAS Log after the macro processor has made its substitutions. Here is part of the SAS Log containing the macro-generated statements when the macro is called with these arguments:

```
MPRINT(HIGHLOW):    title "7 Highest and Lowest Values for HR";
MPRINT(HIGHLOW):    data _null_;
MPRINT(HIGHLOW):    set Tmp(obs=7) Tmp(firstobs=94);
MPRINT(HIGHLOW):    file print;
MPRINT(HIGHLOW):    if _n_ le 7 then do;
MPRINT(HIGHLOW):    if _n_ = 1 then put / "7 Lowest Values";
MPRINT(HIGHLOW):    put "Patno = " Patno @15 "Value = " HR;
MPRINT(HIGHLOW):    end;
MPRINT(HIGHLOW):    else if _n_ ge 8 then do;
MPRINT(HIGHLOW):    if _n_ = 8 then put / "7 Highest Values";
MPRINT(HIGHLOW):    put "Patno = " Patno @15 "Value = " HR;
MPRINT(HIGHLOW):    end;
MPRINT(HIGHLOW):    run;
```

Looking at this section of code, you can see how the macro variables have been replaced by their respective calling arguments. By the way, the missing semicolon at the end of the line where the macro is called is not a mistake—you don't need or want it. The reason is that this macro code contains SAS statements that already end in semicolons. Some macros might only generate a portion of a SAS statement, and a semicolon in the middle of a SAS statement will cause an error.

Because the program has already been described, understanding the macro version is straightforward. All that was necessary to turn the program into a macro was to make the values of the data set name, the variable to test, the ID variable name and the number of observations to list into macro calling arguments. By setting $N=10$ in the named parameter list, this macro will, by default, list the top and bottom 10 values if you do not enter a value for N when you call the macro.

Two other features of this macro bear explaining: First, you need to use the %EVAL macro function when you do arithmetic on the macro variable &N. The reason is that macro values are text and you cannot add or subtract numbers from text values. The %EVAL function treats the macro variables used in arithmetic computations as integers. Second, the macro ends with PROC DATASETS. Although the data set Tmp is a temporary data set and will disappear when you close your SAS session, this author feels that macros should clean up after themselves by deleting temporary data sets.

You might want to place this macro into your macro library and use it as an alternative to PROC UNIVARIATE to list the *n* highest and lowest values of a numeric variable in a data set.

If you call the macro like this:

```
%highlow(Dsn=Clean.Patients, Var=HR, Idvar=Patno, n=7)
```

you will generate the following output:

Figure 4.9: Output from Running the %HighLow Macro with *N*=7

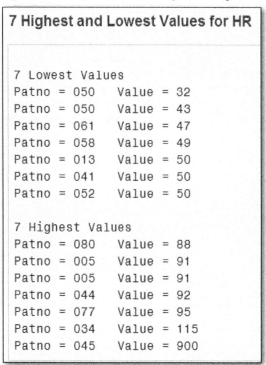

```
7 Highest and Lowest Values for HR


7 Lowest Values
Patno = 050    Value = 32
Patno = 050    Value = 43
Patno = 061    Value = 47
Patno = 058    Value = 49
Patno = 013    Value = 50
Patno = 041    Value = 50
Patno = 052    Value = 50

7 Highest Values
Patno = 080    Value = 88
Patno = 005    Value = 91
Patno = 005    Value = 91
Patno = 044    Value = 92
Patno = 077    Value = 95
Patno = 034    Value = 115
Patno = 045    Value = 900
```

Describing Two Programs to List the Highest and Lowest Values by Percentage

You have seen several ways to list the top and bottom *n* values of a variable. As an alternative, you might want to see the top and bottom *n* percentage of data values. Two very different approaches to this problem are presented in the next two sections.

Using PROC UNIVARIATE

One approach uses PROC UNIVARIATE to output the cutoff values for the desired percentiles—the other uses PROC RANK to divide the data set into groups. Here is the UNIVARIATE approach:

Program 4.5: Listing the Top and Bottom *n*% for HR (Using PROC UNIVARIATE)

```
*Program HighLowPercent
 Prints the top and bottom 5% of values of a variable;

proc univariate data=Clean.Patients noprint; ❶
   var HR;
   id Patno;
   output out=Tmp pctlpts=5 95 pctlpre = Percent_; ❷
run;

data HighLowPercent;
   set Clean.Patients(keep=Patno HR); ❸
   ***Bring in upper and lower cutoffs for variable;
   if _n_ = 1 then set Tmp; ❹
   if HR le Percent_5 and not missing(HR) then do; ❺
      Range = 'Low ';
      output;
   end;
   else if HR ge Percent_95 then do; ❻
      Range = 'High';
      output;
   end;
run;

proc sort data=HighLowPercent; ❼
   by HR;
run;

title "Top and Bottom 5% for Variable HR";
proc print data=HighLowPercent; ❽
   id Patno;
   var Range HR;
run;
```

❶ You are using PROC UNIVARIATE to compute the value of HR at the 5th and 95th percentiles. Because you only want an output data set and you do not want printed output, use the NOPRINT option.

❷ Use an OUTPUT statement to have PROC UNIVARIATE create an output data set. You name this data set Tmp (this author's favorite temporary data set name) and use the keywords PCTLPTS (percentile points) and PCTLPRE (percentile prefix) to name the variables in the output data set that contains the value of HR at the 5th and 95th percentiles. Here's how this works. The numbers you place after PCTLPTS are the percentiles you want to output. You need to supply a name for the two percentiles that you are outputting. They can't be called 5 and 95 because SAS variable names must begin with a letter or underscore. The value you select following PCTLPRE becomes a prefix for the variable names in the output data set. Because the prefix you chose was Percent_, the two variables in data set Tmp will be named Percent_5 and Percent_95.

❸ You need to add the two variables Percent_5 and Percent_95 to each observation in the Patients data set. First, use a SET statement to read each observation in the Patients data set.

❹ This statement is known as a *conditional* SET statement. On the first iteration of the DATA step, _N_ is equal to 1, the IF statement is true, and the SET statement brings the values of Percent_5 and Percent_95 into the PDV. On the next iteration of the DATA step, the second observation is read from the Patients data set. However, the value of _N_ is now 2 and the IF statement is not true. However, because the two values Percent_5 and Percent_95 came from a SAS data set, these values are not set back to missing, as they would be if you were reading raw data—they are retained. Thus, the values of Percent_5 and Percent_95 are added to every observation of the Patients data set.

❺ You can now test if the value of HR is less than the value at the 5th percentile and not missing. If so, you set Range equal to 'Low' (with an extra space so that the length of Range is set to 4) and output an observation.

❻ If the previous comparison is not true, you test if the value of HR is greater than the value at the 95th percentile. If so, you set Range to 'High' and output the observation.

❼ Sort the data set containing only the values of HR below the 5th percentile (and not missing) or above the value of HR at the 95th percentile.

❽ Use PROC PRINT to list the values of HR below the 5th percentile or above the 95th percentile.

Listed below is the output from running Program 4.5:

Figure 4.10: Output from Program 4.5

Top and Bottom 5% for Variable HR

Patno	Range	HR
050	Low	32
050	Low	43
061	Low	47
058	Low	49
013	Low	50
041	Low	50
052	Low	50
066	Low	50
005	High	91
005	High	91
044	High	92
077	High	95
034	High	115
045	High	900

Presenting a Macro to List the Highest and Lowest *n*% Values

Once again, it is a straightforward process to convert Program 4.5 to a macro, as seen in Program 4.6:

Program 4.6: Converting Program 4.5 to a Macro

```
*-----------------------------------------------------------*
| Program Name: HighLowPcnt.sas                             |
| Purpose: To list the n percent highest and lowest values for |
|          a selected variable.                             |
| Arguments: Dsn     - Data set name                        |
|            Var     - Numeric variable to test             |
|            Percent - Upper and Lower percentile cutoff    |
|            Idvar   - ID variable to print in the report   |
| Example: %HighLowPcnt(Dsn=clean.patients,                 |
|                     Var=SBP,                              |
|                     Percent=5,                            |
|                     Idvar=Patno)                          |
*-----------------------------------------------------------*;

%macro HighLowPcnt(Dsn=,   /* Data set name                   */
                Var=,      /* Variable to test                */
                Percent=, /* Upper and lower percentile cutoff */
                Idvar=    /* ID variable                     */);

   ***Compute upper percentile cutoff;
   %let Upper = %eval(100 - &Percent);

   proc univariate data=&Dsn noprint;
      var &Var;
      id &Idvar;
      output out=Tmp pctlpts=&Percent &Upper pctlpre = Percent_;
   run;

   data HiLow;
      set &Dsn(keep=&Idvar &Var);
      if _n_ = 1 then set Tmp;
      if &Var le Percent_&Percent and not missing(&Var) then do;
         range = 'Low ';
         output;
      end;
      else if &Var ge Percent_&Upper then do;
         range = 'High';
         output;
      end;
   run;

   proc sort data=HiLow;
      by &Var;
   run;

   title "Highest and Lowest &Percent% for Variable &var";
   proc print data=HiLow;
      id &Idvar;
      var Range &Var;
   run;
```

```
    proc datasets library=work nolist;
      delete Tmp HiLow;
    run;
    quit;

%mend HighLowPcnt;
```

Writing this macro uses the same process used to create the previous macro. Macro variables are used to replace values for the data set name, the variable to be analyzed, the ID variable, and the lower percentile value. The top percentile value is computed by subtracting the lower percentile value from 100.

Calling this macro like this:

```
%HighLowPcnt(Dsn=Clean.Patients,
             Var=SBP,
             Percent=5,
             Idvar=Patno)
```

produced the following listing:

Figure 4.11: Running the %HighLowPctn Macro

Highest and Lowest 5% for Variable SBP

Patno	range	SBP
003	Low	56
016	Low	92
038	Low	94
074	Low	98
089	Low	98
083	Low	100
029	High	148
060	High	148
066	High	152
013	High	160
019	High	210
023	High	300

Using PROC RANK

An interesting alternative to using PROC UNIVARIATE takes advantage of PROC RANK with an option called GROUPS=. Before we start, let's review PROC RANK:

Suppose you have a data set called Rank_Example, listed below:

Figure 4.12: Listing of Data Set Rank_Example

ID	X	Y
\multicolumn{3}{l}{Listing of Data Set Rank_Example}		
001	5	10
002	3	15
003	7	11
004	2	13
005	.	14

Next, run the following program:

Program 4.7: Demonstrating PROC RANK

```
proc rank data=Rank_Example out=Ranked_Data;
   var X;
   ranks Rank_of_X;
run;

title "Listing of Ranked_Data";
proc print data=Ranked_Data noobs;
run;
```

Now, let's look at the output data set:

Figure 4.13: Listing of Data Set Ranked_Data

ID	X	Y	Rank_of_X
\multicolumn{4}{l}{Listing of Ranked_Data}			
001	5	10	3
002	3	15	2
003	7	11	4
004	2	13	1
005	.	14	.

First of all, what are ranks? As you can see in the listing above, the smallest value of X (2) is given a rank of 1; the next smallest value of X (3) is given a rank of 2, and so forth. Missing values are not assigned a rank. The Rank_of_X values tell you the order, or ranking, of the Xs. You can list as many variables as you wish from the input data set on the VAR statement and make up names for the corresponding ranked variables on the RANKS statement.

> **Beware!** If you do not supply a RANKS statement, PROC RANK will **replace** your original data values with their ranks. You should always list as many variables on the RANKS statement as there are on the VAR statement.

Notice that the output data set Ranked_Data also contains the variables ID and Y. Why? Because PROC RANK is designed to include all the variables from the input data set in the output data set. If you do not want all of the variables from the input data set included in the output data set, use a KEEP= data set option on the input data set. For example, if you only want ID, X, and Rank_of_X in the output data set, write your PROC RANK statement like this:

```
proc rank data=Rank_Example (keep=ID X) out=Ranked_Data;
```

Why is this procedure useful? Sometimes statisticians use it to rank data that might be highly skewed (such as income) and then perform an analysis on the ranks rather than on the original data values. However, you will see in a moment that a procedure option called GROUPS= is going to make PROC RANK one of your favorite go-to procedures. Let's add the option GROUPS=2 to Program 4. 7 and see what happens:

Program 4.8: Adding the GROUPS= Option to PROC RANK

```
proc rank data=Rank_Example out=Ranked_Data groups=2;
   var X;
   ranks Rank_of_X;
run;

title "Listing of Ranked_Data with GROUPS=2";
proc print data=Ranked_Data noobs;
run;
```

Here is the result:

Figure 4.14: Showing the Result of the GROUPS=2 Option

Listing of Ranked_Data with GROUPS=2

ID	X	Y	Rank_of_X
001	5	10	1
002	3	15	0
003	7	11	1
004	2	13	0
005	.	14	.

PROC RANK attempts to place your data into the number of groups you specify. In Figure 4.14, the two observations with the smallest X values are placed into group 0 and the two observations with the highest values of X are placed into group 1. For some unknown reason, when you use the GROUPS= option, the group numbers start from 0 instead of 1. This is annoying, but we have learned to live with it.

If the number of observations in the input data set is a multiple of the number of groups, each group will have the same number of observations (unless there are some repeated values at a group boundary). If this is not the case, the number of observations in each group will not be the same. For example, if you had one more observation in the input data set (X=4, for example), that observation would wind up in group 1 (the higher group). This will not be an issue for us in the program that follows.

Here is the question: How can you use PROC RANK to list the top and bottom 5% of the data values? How many groups do you need so that there are approximately 5% of the observations in each group? You got it— 20 groups. All you need to do is use PROC RANK with the number of groups equal to 100 divided by the percentage in each group. This brings us to Program 4.9:

Program 4.9: Listing the Top and Bottom *n*% for SBP (Using PROC RANK)

```
proc rank data=Clean.Patients(keep=Patno SBP) out=HiLow groups=20;
   var SBP;
   ranks Range;
run;

proc sort data=HiLow(where=(Range in (0,19)));
   by SBP;
run;

proc format;
   value rank 0 = 'Low'
             19 = 'High';
run;

title "Top and Bottom 5% for Variable SBP";
proc print data=HiLow;
   id Patno;
   var Range SBP;
   format Range rank.;
run;
```

You run PROC RANK with `GROUPS=20`. Remember that this results in values of Range (the variable listed on the RANKS statement) of 0 to 19. Next, notice that PROC SORT is doing double duty—first, by including a WHERE= data set option, you are selecting only observations where Range is equal to 0 (lowest 5%) or Range is equal to 19 (highest 5%). It is also sorting the Range data set by HR so that all the low values are listed first, followed by all the high values. You use PROC FORMAT to create the Rank format that assigns the label 'Low' for Range=0 and 'High' for Range=19. Listed below is the output from this program:

Figure 4.15: Output from Program 4.9

Top and Bottom 5% for Variable SBP

Patno	Range	SBP
003	Low	56
016	Low	92
038	Low	94
074	Low	98
089	Low	98
066	High	152
013	High	160
019	High	210
023	High	300

Because of the way PROC RANK groups observations, this list is similar but not identical to the one produced by PROC UNIVARIATE. The differences are mainly due to multiple subjects having the same value of SBP. It is worthwhile to understand PROC RANK, but the bottom line is that using the PROC UNIVARIATE approach might be slightly better. That is why the macro for computing the top and bottom by percentage used that approach instead of the PROC RANK method.

Using Pre-Determined Ranges to Check for Possible Data Errors

You can define reasonable ranges for certain numeric variables. For example, in healthy adults, you would expect a resting heart rate to be somewhere between 40 and 100. Although it is possible for a patient to have a heart rate outside this range, it would be wise to check if that value was an error. Valid ranges for systolic blood pressure would be 50 to 240 and for diastolic blood pressure, 35 to 130. (Note: These are just suggested values; you might find other values more suitable.)

You can write a DATA step to check for out-of-range values as demonstrated in Program 4.10:

Program 4.10: Listing Out-of-Range Values Using a DATA Step

```
*Program to List Out-of-Range Values;
title "Listing of Out-of-Range Values";
data _null_;
   file print;
   set Clean.Patients(keep=Patno HR SBP DBP);
   *Check HR;
   if (HR lt 40 and not missing(HR)) or HR gt 100 then
      put Patno= HR=;
   *Check SBP;
   if (SBP lt 50 and not missing (SBP)) or SBP gt 240 then
      put Patno= SBP=;
```

```
    *Check DBP;
    if (DBP lt 35 and not missing (DBP)) or DBP gt 130 then
       put Patno= DBP=;
run;
```

Once again, you use DATA _NULL_ because you do not need a SAS data set when you are finished looking for possible data errors. Notice that the input data set Clean.Patients uses a KEEP= data set option. As mentioned previously, this makes the program more efficient by bring into the PDV only those variables you need. When you are testing for low values of HR, SBP, or DBP, you do not want to see missing values (at least in this example), so you use the MISSING function to test for this condition.

Here is the output:

Figure 4.16: Output from Program 4.10

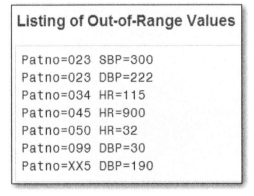

Listing of Out-of-Range Values

```
Patno=023 SBP=300
Patno=023 DBP=222
Patno=034 HR=115
Patno=045 HR=900
Patno=050 HR=32
Patno=099 DBP=30
Patno=XX5 DBP=190
```

Identifying Invalid Values versus Missing Values

If you have invalid data, perhaps a number with two decimal points or a non-digit in the raw data field you are reading, SAS will print an error message to the SAS Log and set the value of the numeric variable to missing. One way to identify these errors is to inspect the SAS Log when you are creating a SAS data set for the first time. This is fine, but SAS stops printing these types of error messages after a predefined number of data errors (usually 20). You can increase the number of errors to report in the SAS Log by using the system option ERRORS=*n*, where *n* is the number of errors to report. If you see a message in the SAS Log that you have exceeded the error reporting limit, rerun the program with ERRORS= set to a larger value. To set the error limit to 40, use the following statement before your DATA statement:

```
options errors=40;
```

One other alternative is to read a numeric data field as character data and use the NOTDIGIT function to test if there are any non-digits in that field. If so, you can print out the invalid values. Here is an example:

Program 4.11: Identifying Invalid Numeric Data

```
*Program to Demonstrate How to Identify Invalid Data;
title "Listing of Invalid Data for HR, SBP, and DBP";
data _null_;
   file print;
```

```
     input @1  Patno $3.
           @4   HR $3.
           @7   SBP $3.
           @10 DBP $3.;
   if notdigit(trimn(HR)) and not missing(HR) then
      put "Invalid value " HR "for HR in patient " Patno;
   if notdigit(trimn(SBP)) and not missing(SBP) then
      put "Invalid value " SBP "for SBP in patient " Patno;
   if notdigit(trimn(DBP)) and not missing(DBP) then
      put "Invalid value " DBP "for DBP in patient " Patno;
datalines;
001080140 90
0029.0180 90
003abcdefghi
00490x120100
005       80
;
```

Here is the output:

Figure 4.17: Output from Program 4.11

```
Listing of Invalid Data for HR, SBP, and DBP

Invalid value 9.0 for HR in patient 002
Invalid value abc for HR in patient 003
Invalid value def for SBP in patient 003
Invalid value ghi for DBP in patient 003
Invalid value 90x for HR in patient 004
```

What if you have numeric values that include decimal points? The program above will flag those values as errors because a period is not a digit. If some of your numeric values contain decimal points, you will need to modify Program 4.11. The following program uses an interesting technique to detect invalid numeric values.

Program 4.12: An Alternative to Program 4.11

```
*Program to Demonstrate How to Identify Invalid Data;
title "Listing of Invalid Data for HR, SBP, and DBP";
data _null_;
   file print;
   input @1  Patno $3.
         @4   HR $3.
         @7   SBP $3.
         @10 DBP $3.;
   X = input(HR,3.);
   if _error_ then do;
      put "Invalid value " HR "for HR in patient " Patno;
      _error_ = 0;
   end;
   X = input(SBP,3.);
```

```
    if _error_ then do;
       put "Invalid value " SBP "for SBP in patient " Patno;
       _error_ = 0;
    end;
    X = input(DBP,3.);
    if _error_ then do;
       put "Invalid value " DBP "for DBP in patient " Patno;
       _error_ = 0;
    end;
datalines;
001080140 90
0029.0180 90
003abcdefghi
00490x120100
005       80
;
```

This program lets SAS do all the work. You read each variable as character and use the INPUT function to attempt a character-to-numeric conversion. When SAS encounters an error reading raw data with an INPUT statement or invalid values with an INPUT function, it sets the internal variable _ERROR_ to 1 (true). This program uses that feature to detect values that are not valid numeric values. Notice that _ERROR_ is set back to 0 after each INPUT function.

Here is the output:

Figure 4.18: Output from Program 4.12

```
Listing of Invalid Data for HR, SBP, and DBP

Invalid value abc for HR in patient 003
Invalid value def for SBP in patient 003
Invalid value ghi for DBP in patient 003
Invalid value 90x for HR in patient 004
```

Notice that the value for patient 002 (that contains a decimal point) is no longer listed as invalid.

If you only have a few errors, it is probably best just to look at the SAS Log (overwriting the default value of 20 errors if necessary). If you have large volumes of data that could have invalid values, you might opt for either of the two programs described here.

Checking Ranges for Several Variables and Generating a Single Report

Most of the programs and macros you have seen so far test out-of-range values for individual variables. The two macros described here will test several variables for out-of-range values and produce a consolidated report. To make the macro more flexible, you can decide to treat missing values for each variable as valid or invalid. The macro is listed first, followed by a step-by-step explanation:

Program 4.13: Checking Ranges for Several Variables

```
*Program Name: Errors.Sas
 Purpose: Accumulates errors for numeric variables in a SAS
          data set for later reporting/
          This macro can be called several times with a
          different variable each time. The resulting errors
          are accumulated in a temporary SAS data set called
          errors.

*Macro variables Dsn and IDvar are set with %Let statements before
 the macro is run;

%macro Errors(Var=,      /* Variable to test      */
              Low=,      /* Low value             */
              High=,     /* High value            */
              Missing=ignore
                         /* How to treat missing values      */
                         /* Ignore is the default. To flag    */
                         /* missing values as errors set      */
                         /* Missing=error                     */);

data Tmp;
   set &Dsn(keep=&Idvar &Var);
   length Reason $ 10 Variable $ 32;
   Variable = "&Var";
   Value = &Var;
   if &Var lt &Low and not missing(&Var) then do;
      Reason='Low';
      output;
   end;
   %if %upcase(&Missing) ne IGNORE %then %do;
   else if missing(&Var) then do;
      Reason='Missing';
      output;
   end;
   %end;

   else if &Var gt &High then do;
       Reason='High';
      output;
      end;
      drop &Var;
   run;

   proc append base=errors data=Tmp;
   run;

%mend errors;
```

```
%macro report;
   proc sort data=Errors;
      by &Idvar;
   run;

   proc print data=errors;
   title "Error Report for Data Set &Dsn";
      id &Idvar;
      var Variable Value Reason;
   run;

   proc datasets library=work nolist;
      delete errors;
      delete tmp;
   run;
   quit;

%mend report;
```

To avoid having to enter the data set name and the ID variable each time this macro is called, the two macro variables &Dsn and &Idvar are assigned with %LET statements. Calling arguments to the macro are the name of the numeric variable to be tested, the lower and upper valid values for this variable, and an indicator to determine if missing values should be listed in the error report. To keep the macro somewhat efficient, only the variable in question and the ID variable are added to the Tmp data set because of the KEEP= data set option. The variables Reason and Variable hold values for why the observation was selected and the name of the variable being tested. Because the name of the numeric variable to be tested changes each time the macro is called, a variable called Value is assigned the value of the numeric variable.

The range is first checked, ignoring missing values. If the value of the macro variable &Missing is not equal to 'IGNORE,' then an additional check is made to see if the value is missing. Finally, each error found is added to the temporary data set Errors by using PROC APPEND. This is the most efficient method of adding observations to an existing SAS data set. Each time the %Errors macro is called, all the invalid observations will be added to the Errors data set.

The second macro, %Report, is a macro that should be called once after the %Errors macro has been called for each of the desired numeric variable range checks. The %Report macro is simple. It sorts the Errors data set by the ID variable, so that all errors for a particular ID will be grouped together. Finally, as you have done in the past, it uses PROC DATASETS to clean up the Work data sets that were created.

To demonstrate how these two macros work, the %Errors macro is called three times, for the variables heart rate (HR), systolic blood pressure (SBP), and diastolic blood pressure (DBP), respectively. For the HR variable, you want missing values to appear in the error report; for the other two variables, you do not want missing values listed as errors. Here is the calling sequence:

```
***Set two macro variables;
%let Dsn=Clean.Patients;
%let IDvar = Patno;

%Errors(Var=HR, Low=40, High=100, Missing=error)
%Errors(Var=SBP, Low=50, High=240, Missing=ignore)
%Errors(Var=DBP, Low=35, High=130)
```

```
***Generate the report;
%report
```

Executing the statements above produces the listing below:

Figure 4.19: Error Report Produced by the Macros %Errors and %Report

Error Report for Data Set Clean.Patients

Patno	Variable	Value	Reason
023	HR	.	Missing
023	SBP	300	High
023	DBP	222	High
034	HR	115	High
045	HR	900	High
050	HR	32	Low
099	DBP	30	Low
XX5	DBP	190	High

The missing value for HR was included in the output because of the `Missing=Error` calling argument. Missing values for DBP are not listed because a default value of IGNORE was set for the macro variable &Missing. The clear advantage of using this macro is that it provides a report that lists all of the out-of-range or missing value errors for each patient, all in one place.

Conclusions

One of the first steps in ensuring accurate data is to use PROC UNIVARIATE (with or without the NEXTROBS= option) and look at the highest and lowest data values. Graphical techniques are also very useful at this stage. Once you have completed this task, the next step depends on whether you can define reasonable ranges for your variables. If so, consider the %Errors and %Report macros to list possible data errors. If ranges are not reasonable for some variables, proceed on to the next chapter to explore methods that use the distribution of data values to automatically identify possible data errors.

Chapter 5: Automatic Outlier Detection for Numeric Data

Introduction

As you saw in the previous chapter, there are many variables where it is possible to specify a reasonable range for numeric values. When this is not possible, there are other tools in the data cleaning toolbox that you can use. Many of these methods look at the distribution of data values and identify values that appear to be outliers. (Note: In epidemiology, there are certain variables such as age or weight where you might have outright liars.)

Automatic Outlier Detection (Using Means and Standard Deviations)

If your data values have a distribution that looks similar to a normal distribution or at least is somewhat symmetrical (as determined by statistical techniques such as computing skewness and kurtosis or inspection of a histogram of the data), you might consider using properties of the distribution to help identify possible data errors. For example, you could decide to flag all values more than two standard deviations from the mean. However, if you had some severe data errors, the standard deviation could be so badly inflated that obviously incorrect data values might lie within two standard deviations of the mean (and not be identified as possible errors).

A possible workaround for this would be to compute the standard deviation after removing some of the highest and lowest values. For example, you could compute a standard deviation of the middle 80% of your data and use this to decide on outliers. Another popular alternative is to use an algorithm based on the interquartile range (the difference between the 25th percentile and the 75th percentile).

Let's first see how you could identify data values more than two standard deviations from the mean. You can use PROC MEANS to compute the mean and standard deviation, followed by a short DATA step to select the outliers, as shown in Program 5.1.

Program 5.1: Detecting Outliers Based on the Standard Deviation

```
*Use PROC MEANS to Output means and standard deviations to a data set;
proc means data=Clean.Patients noprint;
   var HR;
   output out=Mean_Std(drop=_type_ _freq_)
          mean=
          std= / autoname;
run;
```

Before we look at the next part of this program, let's review several features used in the first part. The NOPRINT procedure option has been used before—it suppresses printed output, Remember, you only want PROC MEANS to create a data set containing the mean and standard deviation for heart rate.

You name the output data set Mean_Std following the keyword OUT= on the OUTPUT statement. When PROC MEANS produces output data sets, it adds two variables, _TYPE_ and _FREQ_. Because you don't need these variables for this program, you use a DROP= data set option to drop them. Notice that there are no variable names following the keywords MEAN= and STD=. Rather than make up names for these variables, you are letting PROC MEANS name them for you by including the AUTONAME option on the OUTPUT statement. This extremely useful option creates variable names for all of the output statistics by combining the name of the analysis variable (HR in this example), adding an underscore, followed by the name of the requested statistic (Mean or Stddev in this example—Std is an abbreviation for Stddev). To make this clear, look at a listing of data set Mean_Std below:

Figure 5.1: Listing of Output Data Set Mean_Std

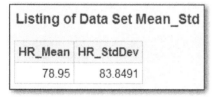

HR_Mean	HR_StdDev
78.95	83.8491

Notice the variable names for the mean and standard deviation in this data set. Using the AUTONAME option with an OUTPUT statement saves you the trouble of naming all the variables and also results in consistent names for these variables. It is recommend that you consider using this option.

Next, take a look at the values for these two variables. A mean of almost 79 for resting heart rate is a bit high. However, the standard deviation of almost 84 is much larger than you would expect. You might recall that there is one data error where the heart rate was entered as 900 (should have been 90). The effect of an outlier like this is to increase the mean somewhat, but the effect on the standard deviation is much more dramatic. If you recall, part of the calculation for a standard deviation is to subtract the mean from each data point and then square the result. That is why extreme outliers have such a major effect on the standard deviation.

Continuing with the program, you need to add the mean and standard deviation to each observation in the original Patients data set. You have already seen a conditional SET statement and we are going to use it here:

Program 5.1 (continued)

```
title "Outliers for HR Based on 2 Standard Deviations";
data _null_;
   file print;
```

```
      set Clean.Patients(keep=Patno HR);
      ***bring in the means and standard deviations;
      if _n_ = 1 then set Mean_Std;
      if HR lt HR_Mean - 2*HR_StdDev and not missing(HR)
         or HR gt HR_Mean + 2*HR_StdDev then put Patno= HR=;
   run;
```

The IF statement checks for all values of HR that are more than two standard deviation from the mean (omitting missing values). The results of running this program on the Patients data set follow:

Figure 5.2: Output from Program 5.1

```
Outliers for HR Based on 2 Standard Deviations

Patno=045 HR=900
```

Well that didn't work very well! The mean heart rate is about 80 and the standard deviation is about 84. Two standard deviations below the mean is a negative number and two standard deviations above the mean is almost 250. That is the reason that only one heart rate (900) was detected by this program.

One way to fix this problem is to compute trimmed statistics. This is done by first removing some values from the top and bottom of the data set, as we demonstrate in the next section.

Detecting Outliers Based on a Trimmed Mean and Standard Deviation

A quick and easy way to compute trimmed statistics and output them to a SAS data set is to first run PROC RANK with the GROUPS= option to divide the data set into *n* groups. For example, if you want to trim 10% from the top and bottom of your data set, you would need to set GROUPS= equal to 10 and remove all the observations for HR in the top and bottom groups. Below is a program to trim 10% off the top and bottom of the heart rate values and compute the mean and standard deviation:

Program 5.2: Computing Trimmed Statistics

```
proc rank data=Clean.Patients(keep=Patno HR) out=Tmp groups=10;
   var HR;
   ranks Rank_HR;
run;

proc means data=Tmp noprint;
   where Rank_HR not in (0,9);
   *Trimming the top and bottom 10%;
   var HR;
   output out=Mean_Std_Trimmed(drop=_type_ _freq_)
         mean=
         std= / autoname;
run;
```

To see exactly what is happening here, first take a look at a portion of the observations in the output data set created by PROC RANK (Tmp):

Figure 5.3: Selected Observations from Data Set Tmp

Selection Observations from Data Set Tmp

Patno	HR	Rank_HR
001	69	4
002	76	6
003	70	4
004	76	6
005	91	9

009	82	8
011	68	3
012	70	4
013	50	0
014	83	8
016	65	2
017	83	8
018	81	8

Remember that when you use the GROUPS= option with PROC RANK, the group numbers start from 0. To trim 10% off the top and bottom of the Patients data set, you need to remove all observations where the value of Rank_HR is 0 or 9.

Adding this statement to PROC MEANS accomplishes the job of trimming the top and bottom 10% of the HR values:

```
where Rank_HR not in (0,9);
```

Here is a listing of the data set Mean_Std_Trimmed:

Figure 5.4: Listing of Data Set Mean_Std_Trimmed

Listing of Data Set Mean_Std_Trimmed

HR_Mean	HR_StdDev
71.0732	7.95943

This is quite a change from the values you computed without any trimming. You can now proceed to run Program 5.1 with the trimmed values. Here is the program:

Program 5.3: Listing Outliers Using Trimmed Statistics

```
title "Outliers for HR Based on Trimmed Statistics";
data _null_;
   file print;
   set Clean.Patients(keep=Patno HR);
   ***bring in the means and standard deviations;
   if _n_ = 1 then set Mean_Std_Trimmed;
   *Adjust the standard deviation;
   Mult = 1.49;
   if HR lt HR_Mean - 2*Mult*HR_StdDev and not missing(HR)
      or HR gt HR_Mean + 2*Mult*HR_StdDev then put Patno= HR=;
run;
```

You are now identifying heart rates outside of two standard deviations of the mean where the mean and standard deviation were computed after trimming the heart rate values by 10%. But wait! What is that variable called Mult doing in the program? Here's the explanation:

Even if you had normally distributed data, when you compute a standard deviation from trimmed data, you obtain a smaller value than if you use all the data values (since the trimmed data has less variation). If you trimmed 10% of the data values from the top and bottom of the data and you had a variable that was normally distributed, your estimate of the standard deviation based on the trimmed data would be too small by a factor of 1.49. If you want to base your decision to reject values beyond two standard deviations, you probably want to adjust the standard deviation you obtained from the trimmed data by that factor.

This value will change depending on how much trimming is done. If you only trim a few percentage points from the top and bottom of the data values, the multiplier will be close to 1 and you can probably ignore it. If you trim a lot (say 25% from the top and bottom), this factor will be larger. The table below shows several trimming values, along with the appropriate MULT factors:

Trim Value (from the top and bottom)	Multiplicative Factor
5%	1.24
10%	1.49
20%	2.12
25%	2.59

The output from Program 5.3 (below) now shows several possible data errors that were not seen when you used untrimmed statistics:

Figure 5.5: Listing of Possible Outliers Based on Trimmed Statistics

```
Outliers for HR Based on Trimmed Statistics

Patno=034 HR=115
Patno=045 HR=900
Patno=050 HR=43
Patno=050 HR=32
Patno=061 HR=47
Patno=077 HR=95
```

Describing a Program that Uses Trimmed Statistics for Multiple Variables

PROC UNIVARIATE has the ability to compute trimmed statistics. However, at the time this book is being written, trimmed statistics cannot be sent to an output data set using an OUTPUT statement (as can be done with means and standard deviations). However, you can still capture the trimmed values and place them in a SAS data set by using the ODS system. Because the program presented here has a number of complexities, it is divided into sections and listings or partial listings of intermediate data sets are displayed so you can see exactly how the program works. If you are simply interested in results, the next section of this chapter presents a macro that you can use, whether or not you understand its inner workings. The first section of the program is shown next:

Program 5.4: Creating Trimmed Statistics Using PROC UNIVARIATE

```
ods output TrimmedMeans=Trimmed;

proc univariate data=Clean.Patients trim=.1;
   var HR SBP DBP;
run;

ods output close;
```

The key to the entire program is revealed in the first line of code. The ODS OUTPUT statement has the following syntax:

   ODS OUTPUT *output-object-name=data-set-name;*

where *output-object-name* is the name of the output object that contains the information you need and *data-set-name* is a data set name that you choose.

The procedure option TRIM= allows you to choose how much to trim from the top and bottom of the data. If you use a decimal value (less than .5), the procedure will treat this value as a proportion—if you use an integer, the procedure will treat this value as the number of observations to trim. You need to submit an ODS OUTPUT CLOSE statement following the procedure (just as you do with any other ODS destination).

The listing below shows the variables that are in the output data set (Trimmed):

Figure 5.6: Listing of Data Set Trimmed

VarName	HalfP	HalfN	Mean	StdMean	LCLMean	UCLMean	DF	tValue	Probt
HR	10.00	10	71.1375	1.229502	68.6902	73.5848	79	57.85878	<.0001
SBP	10.89	11	119.5696	1.253865	117.0734	122.0659	78	95.36082	<.0001
DBP	10.89	11	82.8861	0.971870	80.9512	84.8209	78	85.28518	<.0001

Listing of Data Set Trimmed

The variables you want are the Mean (this is a trimmed mean), the StdMean, and the DF. What you really want is the trimmed standard deviation. However, this data set contains what is called the *standard error* instead. You can compute a standard deviation from a standard error by multiplying the standard error by the square root of the sample size, n. But you don't have n in the data set. No worries. The degrees of freedom (abbreviated DF) is equal to $n - 1$. Therefore, n is equal to DF + 1. By the way, the standard error reported by PROC UNIVARIATE is already adjusted for the amount of trimming you request.

The next problem you need to tackle is to restructure data set Patients so that you can merge it with data set Trimmed. Thus, the next block of code creates a separate observation for each patient and measure (HR, SBP, and DBP):

Program 5.4 (continued)

```
data Restructure;
   set Clean.Patients;
   length VarName $ 32;
   array Vars[*] HR SBP DBP;
   do i = 1 to dim(Vars);
      VarName = vname(Vars[i]);
      Value = Vars[i];
      output;
   end;
   keep Patno VarName Value;
run;
```

This section of code restructures the Patients data set so that it shares the variable VarName with data set Trimmed, thus allowing you to merge the two data sets using VarName as the BY variable. One of the keys to this section of code is the VNAME function. This function takes an array element as its argument and returns the variable name associated with this array element. Here is a listing of the first 10 observations in data set Restructure:

Figure 5.7: First 10 Observations in Data Set Restructure

Patno	VarName	Value
	HR	87
	SBP	128
	DBP	98
001	HR	69
001	SBP	124
001	DBP	86
002	HR	76
002	SBP	130
002	DBP	80
003	HR	70

Listing of RESTRUCTURE (first 10 obs)

You can now sort both data sets (Trimmed and Restructure) by VarName and then merge the two data sets by VarName. The section of code below performs these operations:

Program 5.4 (continued)

```
proc sort data=Trimmed;
   by VarName;
run;

proc sort data=Restructure;
   by VarName;
run;

data Outliers;
   merge Restructure Trimmed;
   by VarName;
   Std = StdMean*sqrt(DF + 1);

   if Value lt Mean - 2*Std and not
   missing(Value) then do;
      Reason = 'Low  ';
      output;
   end;

   else if Value gt Mean + 2*Std then do;
      Reason = 'High';
      output;
   end;
run;
```

All that is left to do is to sort data set Outliers by Patno so that the list of outliers will be in patient number order. Here is the final (yeah!) piece of code:

Program 5.4 (continued)

```
proc sort data=Outliers;
   by Patno;
run;

title "Outliers based on trimmed Statistics";
proc print data=outliers;
   id patno;
   var Varname Value Reason;
run;
```

Here is a partial listing of the final report:

Figure 5.8: Partial Listing of Data Set Outliers

Outliers based on trimmed Statistics

Patno	VarName	Value	Reason
003	SBP	56	Low
007	DBP	102	High
009	DBP	64	Low
013	SBP	160	High
016	SBP	92	Low
019	SBP	210	High
023	DBP	222	High
023	SBP	300	High
029	SBP	148	High
034	HR	115	High
038	SBP	94	Low
045	HR	900	High

This listing is based on trimmed statistics (10% trim) and a cutoff of two standard deviations. The next section presents a macro based on this program. This macro allows you to select the amount to trim and the number of standard deviations for the cutoff.

Presenting a Macro Based on Trimmed Statistics

This macro is presented without extensive explanation, mostly because it is based on Program 5.4, which was described in detail. Here is the listing:

Program 5.5: Presenting a Macro Based on Trimmed Statistics

```
*Method using automatic outlier detection;
%macro Auto_Outliers(
   Dsn=,       /* Data set name                      */
   ID=,        /* Name of ID variable                */
   Var_list=,  /* List of variables to check         */
               /* separate names with spaces         */
   Trim=.1,    /* Integer 0 to n = number to trim    */
               /* from each tail; if between 0 and .5, */
               /* proportion to trim in each tail    */
   N_sd=2      /* Number of standard deviations      */);

   ods listing close;
   ods output TrimmedMeans=Trimmed(keep=VarName Mean Stdmean DF);
   proc univariate data=&Dsn trim=&Trim;
     var &Var_list;
   run;
   ods output close;

   data Restructure;
      set &Dsn;
      length VarName $ 32;
      array Vars[*] &Var_list;
      do i = 1 to dim(Vars);
         VarName = vname(Vars[i]);
         Value = Vars[i];
         output;
      end;
      keep &ID VarName Value;
   run;

   proc sort data=Trimmed;
      by VarName;
   run;

   proc sort data=restructure;
      by VarName;
   run;

   data Outliers;
      merge Restructure Trimmed;
      by VarName;
      Std = StdMean*sqrt(DF + 1);
      if Value lt Mean - &N_sd*Std and not missing(Value)
         then do;
            Reason = 'Low  ';
            output;
         end;
      else if Value gt Mean + &N_sd*Std
         then do;
            Reason = 'High';
            output;
         end;
   run;
```

```
    proc sort data=Outliers;
       by &ID;
    run;

    ods listing;
    title "Outliers Based on Trimmed Statistics";
    proc print data=Outliers;
       id &ID;
       var VarName Value Reason;
    run;

    proc datasets nolist library=work;
       delete Trimmed;
       delete Restructure;
    run;
    quit;
%mend Auto_Outliers;
```

One feature of the macro is that values for Trim (.1) and the number of standard deviations (2) are set as default values. You may ask, "How much should I trim my data?" If you believe that your data set is pretty clean, you might choose a small trim value such as .05 or .1. For data that could contain many errors or where the distributions are heavily skewed, you might need to trim by .2 or even .25. Running this macro as follows results in the same output as in Figure 5.8:

```
%Auto_Outliers(Dsn=Clean.Patients,
               Id=Patno,
               Var_List=HR SBP DBP,
               Trim=.1,
               N_Sd=2)
```

Detecting Outliers Based on the Interquartile Range

Yet another way to look for outliers is a method devised by advocates of exploratory data analysis (EDA). This is a robust method, much like the previous method based on trimmed statistics. It uses the interquartile range (the distance from the 25th percentile to the 75th percentile) and defines an outlier as a multiple of the interquartile range above the 75th percentile or below the 25th percentile. For those not familiar with EDA terminology, the first quartile (Q1) is the value corresponding to the 25th percentile (the value below which 25% of the data values lie). The third quartile (Q3) is the value corresponding to the 75% percentile. For example, you might want to examine any data values more than 1.5 times the interquartile range above Q3 or 1.5 times the interquartile range below Q1 as outliers. This is an attractive method because it is independent of the distribution of the data values.

You can use PROC SGPLOT to display a box plot, a graphical display that uses EDA techniques. As an example, let's generate a box plot for SBP in the Patients data set:

Program 5.6: Using PROC SGPLOT to Create a Box Plot

```
*Using PROC SGPLOT to Create a Box Plot for SBP;
title "Using PROC SGPLOT to Create a Box Plot";
proc sgplot data=clean.Patients(keep=Patno SBP);
   hbox SBP;
run;
```

PROC SGPLOT can create many types of graphical displays such as bar charts, scatter plots, and box plots. To create a box plot, use the HBOX statement followed by the names of one or more variables that you want to display. Here is the output:

Figure 5.9: Box Plot for SBP

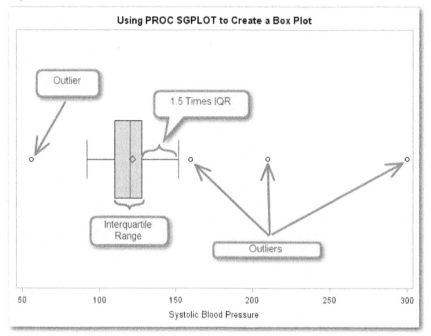

The vertical line in the center of the box is the median and the diamond inside the box represents the mean. The left side and the right side of the box represent the first and third quartiles, respectively. The lines extending from both sides of the box represent a distance of 1.5 times the interquartile range from the sides of the box. (These lines are sometimes referred to as *whiskers* and the box plot is sometimes called a *box-and-whisker plot*.) Finally, the circles represent possible outliers—data points more than 1.5 times the interquartile range below the first quartile or above the third quartile.

You can also use SGPLOT to display box plots at each level of a categorical variable. To demonstrate this, let's create a box plot of SBP for each level of Gender. Because there were a number of data errors for Gender, the program shown next restricts values of Gender to 'F' and 'M':

Program 5.7: Creating a Box Plot for SBP for Each Level of Gender

```
*Using PROC SGPLOT to Create a Box Plot for SBP;
title "Using PROC SGPLOT to Create a Box Plot";
proc sgplot data=clean.Patients(keep=Patno SBP Gender
   where=(Gender in ('F','M')));
   hbox SBP / category=Gender;
run;
```

The output (below) displays a box plot for each value of Gender:

Figure 5.10: Output from Program 5.7

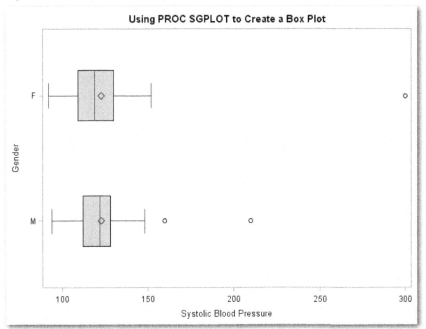

An easy way to determine the interquartile range and the first and third quartiles is to use PROC MEANS to output these quantities. The program below is similar to Program 5.1 except it uses a criterion based on the interquartile range instead of the standard deviation:

Program 5.8: Detecting Outliers Using the Interquartile Range

```
title "Outliers Based on Interquartile Range";
proc means data=Clean.Patients noprint;
   var HR;
   output out=Tmp
          Q1=
          Q3=
          QRange= / autoname;
run;

data _null_;
   file print;
   set Clean.Patients(keep=Patno HR);
   if _n_ = 1 then set Tmp;
   if HR le HR_Q1 - 1.5*HR_QRange and not missing(HR) or
      HR ge HR_Q3 + 1.5*HR_QRange then
      put "Possible Outlier for patient " Patno "Value of HR is " HR;
run;
```

The keywords Q1, Q3, and QRange refer to the first quartile, the third quartile, and the interquartile range, respectively. Here is the output:

Figure 5.11: Output from Program 5.8

Outliers Based on Interquartile Range

```
Possible Outlier for patient 034 Value of HR is 115
Possible Outlier for patient 045 Value of HR is 900
Possible Outlier for patient 050 Value of HR is 32
```

You can adjust Program 5.8 to produce more or fewer possible outliers by changing the number of interquartile ranges from 1.5 to other values.

You might prefer using the interquartile range for heavily skewed data, but with sufficient trimming, you will probably see similar results with the Auto_Outliers macro.

Conclusions

This chapter investigated techniques that use the distribution of data values to identify possible data errors. You saw that using an algorithm based on the standard deviation can fail if there are some extreme values in the data. Ways to overcome this problem included using trimmed statistics or methods based on the interquartile range. There are times when data distributions are sufficiently different from a normal distribution where even trimmed statistics will fail to identify data errors. There is still hope. The next chapter discusses techniques that can be used on highly skewed data sets.

Chapter 6: More Advanced Techniques for Finding Errors in Numeric Data

Introduction

When my wife and I leave our Texas Hill Country home for a trip. we hire a cat-sitter to come every other day to look after our two cats, Dudley and Mickey (you can see their pictures on the back covers of my last two books). I pay her using an online bill paying service. On a recent trip where we were gone longer than normal, I went online to send her a check. When I entered the amount, I received a notice that the amount was much larger than most of my previous checks and questioned whether I was sure that this was the correct amount.

How did they do that? To understand how this and other similar data cleaning tasks work, let's start out with a fictitious data set called Banking.

Introducing the Banking Data Set

One interesting contrast between many types of data (such as data from the banking industry) and health-related data is that many of the physiological values (such as heart rate and blood pressure) conform to predefined ranges. This is not true with banking and other forms of numeric data.

A partial listing of a (fictitious) data set with the types of data used in banking is displayed below:

Figure 6.1: Partial Listing of Data Set Banking

Partial Listing of Data Set Banking

Account	Gender	Balance	Date	Transaction	Deposit	Withdrawal
025461	F	$32,323.80	12/01/2014	1		$743.73
025461	F	$34,205.84	12/20/2014	2	$1,882.04	
025461	F	$33,812.34	01/06/2015	3		$393.50
025461	F	$34,931.60	01/11/2015	4	$1,119.26	
025461	F	$36,606.06	01/27/2015	5	$1,674.46	
025461	F	$35,130.00	02/13/2015	6		$1,476.06
058190	M	$256,309.75	03/08/2015	1	$13,285.95	
058190	M	$268,417.62	03/22/2015	2	$12,107.87	
058190	M	$284,536.20	03/26/2015	3	$16,118.58	
058190	M	$301,661.46	03/27/2015	4	$17,125.26	

Variables in this data set include Account (account number), Gender, Balance (account balance), Transaction (transaction number), Deposit (amount of deposit), and Withdrawal (amount of withdrawal). Some accounts maintain large balances and transactions in these accounts (deposits or withdrawals) might be large. Other accounts have smaller balances and transactions in those accounts will probably be smaller. The techniques for identifying data errors described in the last chapter will not work here. Let's use PROC SGPLOT to examine a histogram of account balances:

Program 6.1: Using PROC SGPLOT to Produce a Histogram of Account Balances

```
title "Histogram of Account Balances";
proc sgplot data=Clean.Banking;
   histogram Balance;
run;
```

You use a HISTOGRAM statement with PROC SGPLOT to generate a histogram, as shown below:

Figure 6.2: Using PROC SGPLOT to Create a Histogram of Account Balances

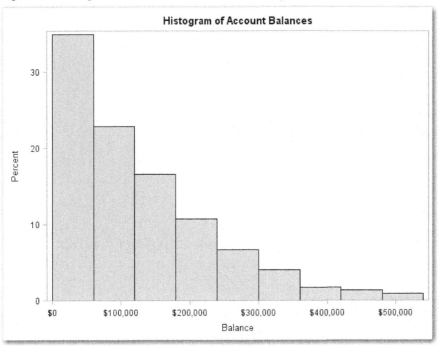

This is a highly skewed distribution. You would expect the distribution of deposits and withdrawals to have similar shapes. To verify this, and to produce some more detailed statistics on deposits, let's see what PROC UNIVARIATE can show us.

Program 6.2: Using PROC UNIVARIATE to Examine Deposits

```
title "Using PROC UNIVARIATE to Examine Bank Deposits";
proc univariate data=Clean.Banking;
   id Account;
   var Deposit;
   histogram / normal;
run;
```

Below are selected portions of the output:

Figure 6.3: Selected Output from Program 6.2

Using PROC UNIVARIATE to Examine Bank Deposits

Variable: Deposit

Moments			
N	156	Sum Weights	156
Mean	8670.61077	Sum Observations	1352615.28
Std Deviation	18152.777	Variance	329523313
Skewness	6.64941276	Kurtosis	51.5473299
Uncorrected SS	6.28041E10	Corrected SS	5.10761E10
Coeff Variation	209.359842	Std Error Mean	1453.38533

Notice the large value for skewness (6.649) showing that the distribution is skewed to the right.

Output from Program 6.2 (continued)

Extreme Observations					
Lowest			Highest		
Value	Account	Obs	Value	Account	Obs
9.66	558805	115	29747.6	058190	21
95.94	558805	112	46011.7	720349	179
106.68	558805	117	94967.4	817303	196
127.53	558805	114	106178.5	647722	135
136.80	649500	142	173371.0	058190	23

There is a wide range of deposit amounts, with the lowest at $9.66 and the highest at $173,371.00. The HISTOGRAM produced by PROC UNIVARIATE shows that deposits are also positively skewed:

Output from Program 6.2 (continued)

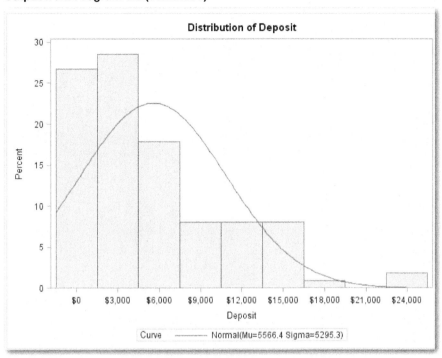

Running the %Auto_Outliers Macro on Bank Deposits

Running a program such as Auto_Outliers (described in the previous chapter) on the variable Deposit will not be very useful. To confirm that, you can submit the following macro call:

Program 6.3: Using the %Auto_Outliers Macro for the Variable Deposit

```
%Auto_Outliers(dsn=Clean.Banking,
               ID=Account,
               Var_List=Deposit,
               Trim=.1,
               N_Sd=2)
```

Here is the result:

Figure 6.4: Output from Program 6.3

Outliers based on trimmed Statistics

Account	Varname	Value	Reason
058190	Deposit	17,125.26	High
058190	Deposit	18,835.52	High
058190	Deposit	18,719.28	High
058190	Deposit	17,608.41	High
058190	Deposit	18,722.05	High
058190	Deposit	29,747.57	High
058190	Deposit	22,013.99	High
058190	Deposit	173,371.00	High
647722	Deposit	106,178.50	High
715555	Deposit	16,562.25	High
715555	Deposit	23,856.74	High
720349	Deposit	46,011.70	High
817303	Deposit	94,967.40	High

Because of the large variation in deposit amounts, it is impossible to tell if any of the values listed in this table represent errors. You need to look at the distribution of deposits within each account if you expect to see an unusually high or low deposit.

Identifying Outliers Within Each Account

One way to look for outliers in the Banking data set is to look within each account. To accomplish this, run PROC MEANS for each account number to create summary data for each account. For this example, the method based on the interquartile range is used to identify possible deposit errors. Trimmed statistics would be an alternative. Use the following program to create a data set containing Q1, Q3, and the interquartile range for each account (the median and sample size (*n*) are requested as well):

Program 6.4: Creating Summary Data for Each Account

```
proc means data=Clean.Banking noprint nway;
   class Account;
   var Deposit;
   output out=By_Account(where=(Deposit_N ge 5) drop=_type_ _freq_)
      Q1= Q3= QRange=
      Median= n= / autoname;
run;
```

The NWAY option is needed to ensure that the output data set only includes summary statistics for each account and not statistics for the entire data set (_TYPE_=0). Notice the two data set options for the output data set By_Account. The WHERE= option restricts accounts with five or more transactions and the DROP= data set option drops the two procedure-generated variables _TYPE_ and _FREQ_. Using a cutoff of five transactions was somewhat arbitrary, but you would not be able to compute reasonable statistics with too few transactions. Using the OUTPUT option AUTONAME automatically generates variable names for the statistics in the output data set. This is how the variables Deposit_Q1, Deposit_Q2, etc., were created. A partial listing of the By_Account data set is shown below:

Figure 6.5: Partial Listing of the By_Account Data Set

Listing of Data Set By_Account

Account	Deposit_Q1	Deposit_Q3	Deposit_QRange	Deposit_Median	Deposit_N
058190	$14,220.81	$18,778.79	$4,557.97	$16,653.78	16

Account	Deposit_Q1	Deposit_Q3	Deposit_QRange	Deposit_Median	Deposit_N
189996	$5,206.40	$7,428.54	$2,222.14	$6,143.69	11

Account	Deposit_Q1	Deposit_Q3	Deposit_QRange	Deposit_Median	Deposit_N
268489	$2,751.49	$3,375.97	$624.48	$3,180.90	10

Account	Deposit_Q1	Deposit_Q3	Deposit_QRange	Deposit_Median	Deposit_N
373920	$6,775.31	$9,483.68	$2,708.37	859.3	

This data set includes Q1, Q3, and QRange for each separate account. You can now run a program using the interquartile range to look for outliers for each account number:

Program 6.5: Looking for Possible Outliers by Account

```
data Outliers;
   merge Clean.Banking(keep=Account Deposit
         where=(Deposit is not missing))
         By_Account(In=In_By_Account);
   by Account;
   if In_By_Account;
   if Deposit lt Deposit_Q1 - 1.5*Deposit_QRange or
      Deposit gt Deposit_Q3 + 1.5*Deposit_QRange then output;
run;
```

Because the Banking data set and the summary data set (By_Account) both contain the variable Account, you can use a MERGE statement to combine the information from these two data sets. To create a nicer looking report, PROC REPORT was used instead of PROC PRINT. Below are the PROC REPORT statements that were used:

Program 6.6: Using PROC REPORT to Create the Listing

```
proc report data=Outliers headline;
   columns Account Deposit Deposit_Median Deposit_QRange;
   define Account / order "Account Number" width=7;
   define Deposit / Format=dollar12.2;
```

```
      define Deposit_Median / "Median" Format=dollar12.2;
      define Deposit_QRange / "Interquartile Range" width=13;
   run;
```

If you are not familiar with PROC REPORT, it is used when you want a bit more control over the appearance of a report. Yes, it can get quite complicated, but the way it is used in this program is not much more complicated than using PROC PRINT. Here are a few salient features of this procedure:

The HEADLINE option on the procedure statement adds a line between the title and the report. Instead of a VAR statement, PROC REPORT uses a COLUMNS statement. You supply a list of variables you want in your report and, like PROC PRINT, the order that you choose for this list is the order of the columns in the report. You typically write a DEFINE statement for every variable in the COLUMNS list. Unlike PROC PRINT, the report you get without explicitly defining how to display each variable might not be very easy to read. You list each variable you want to define following each DEFINE statement and then add options, after the forward slash.

Options used in this program include labels (text in single or double quotes), column widths (WIDTH= option), and formats (FORMAT= option). The ORDER option used in defining the variable Account is especially useful (and not available in PROC PRINT). It says to order the report in account number order. One other difference between PROC REPORT and PROC PRINT is that the default column headings in PROC PRINT are variable names—in PROC REPORT, variable labels (or labels included in your DEFINE statements) are used.

Here is the listing:

Figure 6.6: Listing of Data Set Outliers

Listing of Data Set Outliers

Account Number	Deposit	Median	Interquartile Range
058190	$29,747.57	$16,653.78	$4,557.97
	$173,371.00	$16,653.78	$4,557.97
476397	$2,257.67	$4,490.53	$679.33
494642	$156.94	$1,044.32	$493.94
647722	$106,178.50	$10,657.03	$4,983.40
649500	$136.80	$1,383.30	$423.73
720349	$46,011.70	$3,552.43	$1,880.72
817303	$94,967.40	$8,140.52	$3,402.00

Some of these deposits look quite suspicious. The first deposit for account 058190 seems reasonable, because the median deposit for this account is close to $16,000 and the interquartile range is around $4,500. However, the next deposit of $173,371 certainly bears some investigation. Looking at this listing also raises suspicion about the last three accounts in the list. Remember, you can increase or decrease the sensitivity of this search by decreasing or increasing the multiplier of the interquartile range to flag as possible errors.

Using Box Plots to Inspect Suspicious Deposits

Although you can inspect the list in Figure 6.6 to determine which deposits are suspicious, the task can be made easier by generating box plots for each account in the list. The first step is to create a data set with a list of all the suspicious account numbers. You might be tempted to use the Outliers data set created in Program 6.5; however, there is one account (058190) with two suspicious deposits. You can use PROC SORT with the NODUPKEY option to create a data set with one observation per account number. You can then use this data set to select all the observations from the original Banking data set that might contain errors. The program shown next does this:

Program 6.7: Creating a Data Set of All the Deposits for Suspicious Accounts

```
proc sort data=Outliers(keep=Account) nodupkey out=Single;
   by Account;
run;

Data Plot_Data;
   merge Single(in=In_Single)
         Clean.Banking(keep=Account Deposit
                       where=(Deposit is not missing));
   by Account;
   if In_Single;
run;
```

You use the NODUPKEY option of PROC SORT to create a data set (Single) with one observation for each account number. To help visualize this, here is a listing of data set Single:

Figure 6.7: Listing of Data Set Single

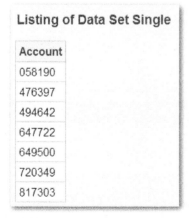

Account
058190
476397
494642
647722
649500
720349
817303

You want to use this data set to select observations from the Banking data set. The key here is to perform the merge on the two data sets (Single and Clean.Banking) using the In_Single temporary variable to select only observations in the Banking data set with an observation in the Single data set. The variable In_Single will be true when there is a contribution to the merge from the Single data set. For those readers more familiar with SQL than the DATA step, you are performing the equivalent of a left join. Because some of the transactions in the Banking data set are withdrawals (where Deposit is a missing value), you also want to delete those observations.

Here is a partial listing of data set Plot_Data:

Figure 6.8: Partial Listing of Data Set Plot_Data

Partial Listing of Data Set Plot_Data

Account	Deposit
058190	$13,285.95
058190	$12,107.87
058190	$16,118.58
058190	$17,125.26
058190	$15,155.67
058190	$11,615.80
058190	$16,182.29
058190	$12,419.71
058190	$18,835.52
058190	$15,434.74
058190	$18,719.28
058190	$17,608.41
058190	$18,722.05
058190	$29,747.57
058190	$22,013.99
058190	$173,371.00
476397	$3,516.29
476397	$4,434.63

It's now time to use this data set to generate box plots for deposits in each account. You can do this two different ways: One way is to create a single box plot showing all the separate account numbers in a single plot; the other way is to use the variable Account as a BY variable and generate a separate plot for each account number. The next program uses the first method:

Program 6.8: Creating a Single Box Plot Showing All Accounts

```
title "Box Plots for Possible Deposit Outliers";
proc sgplot data=Plot_Data;
   hbox Deposit / category=Account;
run;
```

Use the HBOX statement to create a box plot for the variable Deposit using Account as the category variable. Here is the output:

Output from Program 6.8

Because of the large differences in deposits in these accounts, some of the box plots shown here are scrunched up and hard to read. However, you can see there are obvious problems with accounts 058190, 647722, 720349, and 817303.

Let's try the second method (using Account as a BY variable) to get a better picture of these plots. The program looks like this:

Program 6.9: Creating a Separate Box Plot for Each Account

```
title "Creating Separate Plots for Each Account";
proc sgplot data=Plot_Data;
   hbox Deposit;
   by Account;
run;
```

Data set Plot_Data is already sorted by Account, so it is not necessary to sort it again. Below are two of the seven plots:

Figure 6.9: Selected Individual Box Plots (Account 476397)

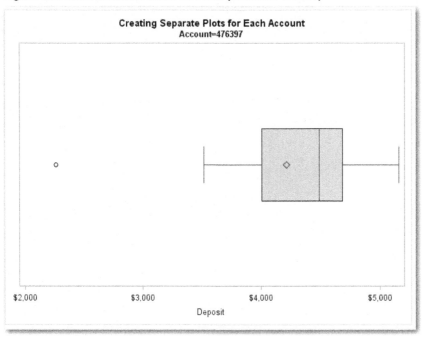

The outlier on this plot does not seem too suspicious—although it did meet the criterion of being more than 1.5 interquartile ranges below the first quartile.

Figure 6.10: Selected Individual Box Plots (Account 817303)

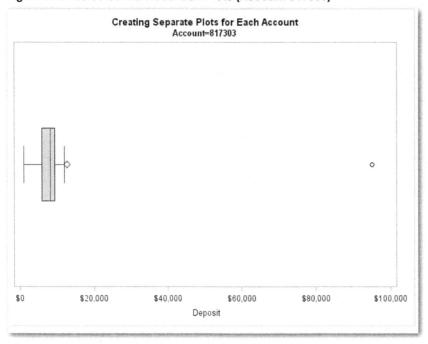

This plot should clearly send up a red flag. You see a deposit close to $100,000 where the median deposit is around $10,000.

Using Regression Techniques to Identify Possible Errors in the Banking Data

You would expect deposits (or withdrawals) to be correlated with the account balance. It would be suspicious to deposit very large amounts into an account with a small balance. It would be even more unusual to withdrawal amounts as large or larger than the account balance.

You should, therefore, be able to run a regression between account balances and deposits and look for outliers. There are a number of ways to detect influential data points when performing simple or multiple regression. Most of these methods involve running the regression with all the data points included and then running it again with each data point removed, looking for differences either in the predicted values or the slopes (coefficients of the independent variables).

Let's use PROC REG to regress Deposit against Balance. Keep in mind that, statistically, this is a bit unorthodox because you are using a data set (Banking) where there are multiple observations per account number. We can still obtain useful information by performing the regression in this manner.

After running the regression, you can use some of the diagnostic plots and tables to attempt to identify influential data points that might be data errors. Here are the PROC REG statements with requests for almost all of the diagnostic plots that are available:

Program 6.10: Using PROC REG to Regress the Account Balance Against Deposits

```
title "Regression of Deposit by Balance";

proc reg data=Clean.Banking(where=(Deposit is not missing)
   keep=Account Deposit Balance)
   plots(only label)=(diagnostics(unpack)
   residuals(unpack)
   rstudentbypredicted dffits fitplot observedbypredicted);
   id Account;
   model Deposit=Balance / influence;
   output out=Diagnostics rstudent=Rstudent cookd=Cook_D
                          dffits=DFfits;
   run;
quit;
```

Because some observations are for withdrawals (where the variable Deposit is missing), you use a WHERE= data set option to include only observations where Deposit is not missing. The KEEP= data set option brings in the variables PROC REG needs to perform the regression.

The PLOTS procedure option allows you to choose which plots you want and whether you want the plots in panels (several plots together on a page) or individual plots (requested by the keyword UNPACK). Key words ONLY and LABEL tell the procedure to only output requested plots and omit plots included in the default list. LABEL will label points on selected plots, using Account (the ID variable). It is not necessary to be a statistician to use much of the information in the output.

It is important to include an ID statement in your procedure, especially when you use the LABEL option—it uses this variable for the labels. In addition to the regression tables and diagnostic plots, there is also a request to place the diagnostic values in an output data set. This is accomplished by the OUTPUT statement (this data set will be used later).

Let's begin with some of the regression tables:

Figure 6.11: Selected Portions of the Regression Statistics

Regression of Deposit by Balance

Model: MODEL1
Dependent Variable: Deposit

Number of Observations Read	156
Number of Observations Used	156

Analysis of Variance

Source	DF	Sum of Squares	Mean Square	F Value	Pr > F
Model	1	13494235009	13494235009	55.30	<.0001
Error	154	37581878481	244038172		
Corrected Total	155	51076113490			

Root MSE	15622	R-Square	0.2642
Dependent Mean	8670.61077	Adj R-Sq	0.2594
Coeff Var	180.16864		

The first step is to look at the number of observations read and the number of observations used (top portion of the table). Because you used a WHERE= data set option to include only those observations with non-missing values of Deposit, the number of observations read and used are equal.

Next, notice the *p*-value. It is highly significant ($p < .0001$). Also notice that the *r*-square is .2642. This is a measure of how much the variance of one variable can explain the variance in another variable. You would expect this to be higher, but remember that there are some extreme data errors that will influence the regression.

Figure 6.12 is a regression plot showing each of the data points (Deposit and Balance), the regression line, and two types of confidence intervals:

Figure 6.12: Displaying the Regression Plot

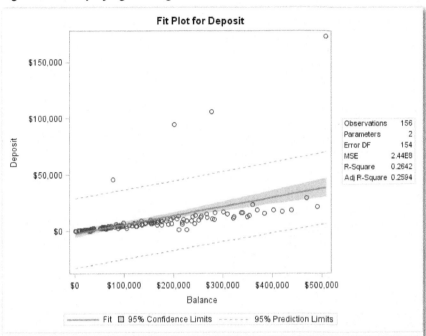

The narrower band, closest to the regression line, tells you that, given a value of Balance, you are 95% confident that the mean Deposit will be within that interval. The wider band (with the dotted lines) tells you that, given a value of Balance, you expect 95% of the individual deposits to lie within that band. The latter interval is what you are more interested in at this point. You can clearly see four data points that do not play well with others. You will see several ways of identifying these accounts where the deposit was much larger than you would expect.

Now let's look at some of the diagnostic plots:

Figure 6.13: Identifying Accounts with Large Residuals

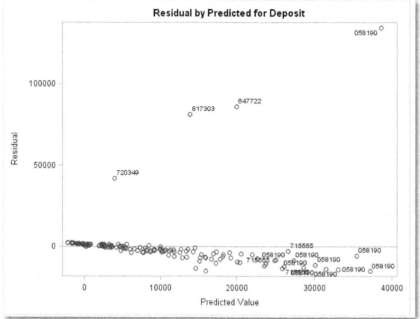

This plot shows the predicted value of Deposit on the X axis and residuals on the Y axis. A *residual* is the difference between the value predicted by the regression equation and the actual value of each data point. Because you included an ID statement in PROC REG and added the LABEL option on the PLOTS request, you see the account numbers listed next to each data point. The four suspicious data points are accounts 720349, 817303, 647722, and 058190. Because it might be difficult to identify suspicious data points from a graph, especially in a large data set, you will see in the next section how to use the output data set that you requested to identify these points.

Two additional measures of influence (data points that have a large effect on the regression) are Cook's D and DFfits. Let's take a look at their plots:

Figure 6.14: Using Cook's D to Identify Influential Points

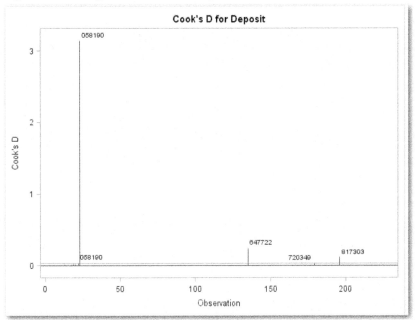

Cook's D measures the effect of each data point (being included or not) on the overall estimates of the parameters (coefficients) in the model. This plot identifies several abnormal values. We'll investigate them more in detail in the next section.

Figure 6.15: Using DFfits to Identify Influential Points

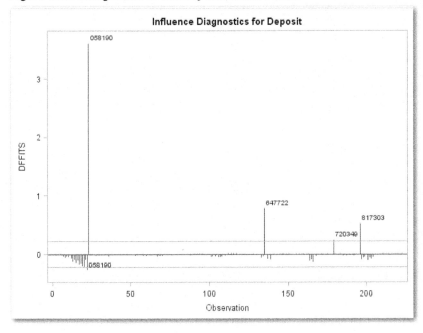

DFfits measures the effect of each data point on the predicted value. In this plot, you see several values that might indicate outliers (or errors).

Using Regression Diagnostics to Identify Outliers

It's fine to look at the diagnostic plots, but it could be difficult to identify possible data errors from them. This section shows you how to use the output data set (Diagnostics) created by PROC REG. This data set contains the account numbers and deposit amounts, along with the diagnostic measures Rstudent, Cook's D, and DFfits. Before we get into the explanation of these diagnostic variables, let's look at the first few observations in the Diagnostics data set listed below:

Figure 6.16: First Few Observations in Data Set Diagnostics

Partial Listing of Data Set Diagnostics

Account	Balance	Deposit	Cook_D	Rstudent	DFfits
025461	$34,205.84	$1,882.04	.000053434	0.09558	0.010304
025461	$34,931.60	$1,119.26	.000010686	0.04288	0.004608
025461	$36,606.06	$1,674.46	.000027947	0.06985	0.007452
058190	$256,309.75	$13,285.95	.000694336	-0.32067	-0.037156
058190	$268,417.62	$12,107.87	.001583944	-0.45951	-0.056140
058190	$284,536.20	$16,118.58	.000701928	-0.28520	-0.037356
058190	$301,661.46	$17,125.26	.000955424	-0.30958	-0.043585
058190	$309,800.0		91	0.47960	

Rstudent is a measure of the residual for each data point—the distance that data point is from the regression line. The value computed by PROC REG is quite clever. It computes the distance of each data point from the regression line computed with that data point removed from the regression calculation. Here is the reason: A strong outlier tends to move the regression line closer to it than if that data point is not included. Therefore, if you measure the distance of an outlier from a regression line that was computed with the outlier included, your estimate of the residual will be too small. Data points with absolute values of Rstudent greater than 2 are often considered influential. If you have very large data sets, you might want to consider a higher cutoff, such as 3, for examination.

Cook's D measures a data point's influence on the coefficients of the regression equation. Values of Cook's D larger than $4/n$, where n is the sample size, are considered influential.

Finally, the diagnostic called DFfits measures the effect of each data point on the predicted value. The regression is run with and without each data point and the difference between the predicted values is used to compute the DFfits statistic. Values of DFfits greater than $2*Sqrt(p/n)$, where p is equal to the number of parameters in the model (the number of independent variables in the model plus the intercept) and n is the sample size, are considered influential.

Let's see how each of these diagnostic values can be used to identify suspicious deposits.

The following piece of SAS code computes the sample size and places that value in a macro variable, so that it can be used later:

Program 6.11: Computing the Sample Size and Placing the Value in a Macro Variable

```
data _null_;
    if 0 then set Diagnostics nobs=Number_of_Obs;
    call symputx('N',Number_of_Obs);
    stop;
run;
```

This is the same technique you used in Chapter 4 (Program 4.3). Briefly (so you don't have to flip back to Chapter 4), the option NOBS= places the number of observations in data set Diagnostics into the variable Number_of_Obs. The CALL SYMPUTX routine places the value of Number_of_Obs into a macro variable called *N*. The STOP statement is necessary to prevent this program from looping.

You can now write a DATA step that uses the appropriate criteria to check each observation for a suspicious deposit. Here is the DATA step:

Program 6.12: Identifying Possible Outliers Using the Three Diagnostic Criteria

```
%let  N_Parameters=2;

data Influence;
    set Diagnostics nobs=N;
    Student=0;
    Cook=0;
    DF_fits=0;
    if abs(Rstudent gt 2) then Student=1;
    if Cook_D gt 4/N then Cook=1;
    if DFfits gt 2*sqrt(&N_Parameters/N) then DF_fits=1;
run;
```

The number of parameters was assigned to a macro variable using a %LET statement. This way, if you have a model with a different number of parameters, you can simply change 2 to the appropriate value. Three variables, Student, Cook, and DF_fits, are initially set to 0. Next, each of the three diagnostic values is checked to see if they exceed a commonly used value for outlier detection. You are now ready to use any one of these logical variables (so called because they have values of 0 or 1) to list suspicious deposits.

Below are the first few observations in data set Influence:

Figure 6.17: First Few Observations in Data Set Influence

Listing of Data Set Influence

Account	Balance	Deposit	Cook_D	Rstudent	DFfits	Student	Cook	DF_fits
025461	$34,205.84	$1,882.04	0.00005	0.0956	0.01030	0	0	0
025461	$34,931.60	$1,119.26	0.00001	0.0429	0.00461	0	0	0
025461	$36,606.06	$1,674.46	0.00003	0.0698	0.00745	0	0	0
058190	$256,309.75	$13,285.95	0.00069	-0.3207	-0.03716	0	0	0
058190	$268,417.62	$12,107.87	0.00158	-0.4595	-0.05614	0	0	0
058190	$284,536.20	$16,118.58	0.00070	-0.2852	-0.03736	0	0	0
058190	$301,661.46	$17,125.26	0.00096	-0.3096	-0.04358	0	0	0

First, let's see how each of these indicators do by themselves. Following that, we will use all three together:

Program 6.13: Listing Outliers Based on Rstudent

```
title "Outliers Based on - Rstudent";
proc print data=Influence;
   where Student;
   id Account;
   var Deposit Balance Rstudent;
run;
```

If you prefer to write the WHERE statement as:

```
where Student=1;
```

that is fine, but because all numerical values in SAS that are not 0 or missing are considered true, you can write the statement as shown in Program 6.13. Here is the listing:

Figure 6.18: Outliers Based on Rstudent

Outliers Based on - Rstudent

Account	Deposit	Balance	Rstudent
058190	$173,371.00	$508,504.55	12.9086
647722	$106,178.50	$277,063.69	6.2047
720349	$46,011.70	$78,212.08	2.7619
817303	$94,967.40	$201,332.94	5.7278

You may notice that these four deposits are the same ones you identified in the diagnostic plots, and all four are also identified using the interquartile criteria described earlier in this chapter.

The code to list outliers based on Cook's D and DFfits is identical to Program 6.13, with the variable Student replaced by Cook or DF_Fits (the other two logical variables).

Here are the other two listings based on Cook's D and DFfits:

Figure 6.19: Outliers Based on Cook's D

Outliers Based on - Cook'D

Account	Deposit	Balance	Cook_D
058190	$22,013.99	$491,167.45	0.03611
058190	$173,371.00	$508,504.55	3.14799
647722	$106,178.50	$277,063.69	0.24906
720349	$46,011.70	$78,212.08	0.02976
817303	$94,967.40	$201,332.94	0.11515

There deposits are identical to the ones identified by Rstudent, with one extra deposit made in account 058190.

Figure 6.20: Outliers Based on DFfits

Outliers Based on - DFfits

Account	Deposit	Balance	DFfits
058190	$173,371.00	$508,504.55	3.61491
647722	$106,178.50	$277,063.69	0.78703
720349	$46,011.70	$78,212.08	0.24914
817303	$94,967.40	$201,332.94	0.52712

These are the same deposits as identified by Rstudent.

For maximum sensitivity, you can print out account numbers and deposits based on any one of these criteria exceeding a given limit. Here is such a program:

Program 6.14: Using All Three Criteria to Identify Outliers

```
title "Outliers Based on - All Measures Together";
proc print data=Influence;
   where Student or Cook or DF_fits;
   id Account;
   var Deposit Balance Rstudent Cook_D DFfits;
run;
```

This program will print out any account number and deposit that violates one or more of the diagnostic criteria. Here is the result:

Figure 6.21: Outliers Based on All Three Criteria

Outliers Based on - All Measures Together

Account	Deposit	Balance	Rstudent	Cook_D	DFfits
058190	$22,013.99	$491,167.45	-1.0054	0.03611	-0.26873
058190	$173,371.00	$508,504.55	12.9086	3.14799	3.61491
647722	$106,178.50	$277,063.69	6.2047	0.24906	0.78703
720349	$46,011.70	$78,212.08	2.7619	0.02976	0.24914
817303	$94,967.40	$201,332.94	5.7278	0.11515	0.52712

These are the same observations flagged using Cook's D. Still, using all three maximizes your chances of identifying outliers.

Conclusions

This chapter deals with numeric data that does not lend itself to data cleaning methods discussed in the previous chapters. The two basic techniques presented here were: 1) If there are enough observations for some BY group (such as Account in these examples) to be used, you can use methods similar to those described in the previous chapter adding a BY (or possible CLASS) statement to the programs. If you have a situation where a number of numeric variables are correlated, you can use regression diagnostics to look for possible outliers.

Chapter 7: Describing Issues Related to Missing and Special Values (Such as 999)

Introduction

Many data sets contain missing values. There are several ways in which missing values can enter a SAS data set. First of all, the raw data value might be missing, either intentionally or accidentally. Next, an invalid raw value can cause a missing SAS value to be created. For example, reading a character value with a numeric informat will generate a missing value. Invalid dates are another common cause of SAS generated missing values. Finally, many operations, such as assignment statements, can create missing values. This chapter investigates ways to detect and count missing values for numeric and character variables as well as ways to identify the location and frequency of special code values, such as 999 or 888, in data files.

Inspecting the SAS Log

It is vitally important to carefully inspect the SAS Log, especially when creating a SAS data set for the first time. A log filled with messages about invalid data values is a clue that something might be wrong, either with the data or the program. If you know that a numeric field contains invalid character values, you could choose to read those data values with a character informat (see Chapter 4). This will keep the SAS Log cleaner and make it easier to spot unexpected errors. Let's look at portions of the SAS Log that were generated when the Patients data set was created.

Figure 7.1: Inspecting the SAS Log After Running Patients.sas

```
NOTE: The infile 'c:\books\clean3\Patients.txt' is:
      Filename=c:\books\clean3\Patients.txt,
      RECFM=V,LRECL=32767,File Size (bytes)=4040,
      Last Modified=07Oct2016:13:24:31,
      Create Time=16Sep2016:09:54:42

NOTE: Invalid data for Visit in line 30 12-21.
RULE:       ----+----1----+----2----+----3----+----4----+----5----+--
30          029NY29028F99/99/9999 79148 88844.7900 38
Patno=029 Account_No=NY29028 Gender=F Visit=. HR=79 SBP=148 DBP=88
Dx=844.790 AE=0 _ERROR_=1 _N_=30
NOTE: Invalid data for Visit in line 57 12-21.
57          056PA88813F10/35/2015 70108 98446.3000 38
Patno=056 Account_No=PA88813 Gender=F Visit=. HR=70 SBP=108 DBP=98
Dx=446.300 AE=0 _ERROR_=1 _N_=57
NOTE: 101 records were read from the infile
      'c:\books\clean3\Patients.txt'.
      The minimum record length was 38.
      The maximum record length was 38.
NOTE: The data set CLEAN.PATIENTS has 101 observations and 9 variables.
NOTE: DATA statement used (Total process time):
      real time            0.07 seconds
      cpu time             0.03 seconds
```

The first step is to check that the correct input file was read. It would be a good idea to look at the date it was last modified so that you can be sure it is the most recent file. Next, notice the first NOTE telling you that you have invalid data for Visit (visit date) in line 30, columns 12-21. SAS then prints the line in question and even places a ruler above the line so that you can identify the columns SAS is trying to read as the visit date. You see 99/99/9999 in the date field. Does the DATA step stop because of a reading error? No, it assigns a missing value for Visit and moves right along.

There is a second invalid date in line 57 (10/35/2015). This date gets the same treatment as the previous invalid date. Finally, make a note of how many records were read from the text file and how many observations and variables are in the newly created SAS data set.

Using PROC MEANS and PROC FREQ to Count Missing Values

There are several procedures that will count missing values for you. It might be normal to have missing values for certain variables in your data set. There could also be variables for which missing values are not permitted (such as a patient ID or an account number).

Counting Missing Values for Numeric Variables

An easy way to count missing values for numeric variables is by using PROC MEANS; for character variables, PROC FREQ will provide this information. Program 7.1 is a simple program that can be used to check the number of numeric missing values in the Patients data set.

Program 7.1: Checking Numeric Missing Values

```
title "Checking Numeric Missing Values from the Patients data set";
proc means data=Clean.Patients n nmiss;
run;
```

You probably don't need to run this program because you previously ran PROC UNIVARIATE on the Patients data set. As you may recall, part of the output from PROC UNIVARIATE is a report on missing values. However, if you need to run a check on missing numeric values, Program 7.1 is quick and easy. Here is the output:

Figure 7.2: Output from Program 7.1

Checking Numeric Missing Values from the Patients data set

Variable	Label	N	N Miss
Visit	Visit Date	98	3
HR	Heart Rate	100	1
SBP	Systolic Blood Pressure	101	0
DBP	Diastolic Blood Pressure	101	0
AE	Adverse Event?	101	0

Inspection of the output shows three missing values for Visit (visit date) and one missing value for HR (heart rate) in this data set. If missing values are not allowed for either of these variables, you will have to determine which patients had these missing values and correct the situation.

Counting Missing Values for Character Variables

What about counting missing values for character variables? You would probably not want to run PROC FREQ on variables such as Patno, especially if you had a very large data set. Program 7.2, below, presents a useful way to count the number of missing and non-missing values for character variables:

Program 7.2: Counting Missing Values for Character Variables

```
title "Checking Missing Character Values";
proc format;
   value $Count_Missing ' '   = 'Missing'
                  other = 'Nonmissing';
run;

proc freq data=Clean.Patients;
   tables _character_ / nocum missing;
   format _character_ $Count_Missing.;
run;
```

You start out by creating a character format (it was called $Count_Missing in this example) that has only two value ranges, one for missing values and the other for everything else. Using this format, you can have PROC FREQ count missing and non-missing values for you.

Notice also that it is necessary to use the SAS keyword _CHARACTER_ in the TABLES statement (or to provide a list of character variables). PROC FREQ can produce frequency tables for numeric as well as character variables. Finally, the TABLES option MISSING is an instruction to include the missing values in the body of the frequency listing.

Note: If you use the MISSING option with PROC FREQ and you request percentages, the percentage calculation uses all the values, missing and non-missing, in the denominator rather than just the number of non-missing values.

Examination of the listing from Program 7.2 is a good first step in your investigation of missing values. The output for this program is shown next:

Figure 7.3: Output from Program 7.2

Checking Missing Character Values

Patient Number		
Patno	Frequency	Percent
Missing	1	0.99
Nonmissing	100	99.01

Account Number		
Account_No	Frequency	Percent
Nonmissing	101	100.00

Gender		
Gender	Frequency	Percent
Missing	3	2.97
Nonmissing	98	97.03

Diagnosis Code		
Dx	Frequency	Percent
Missing	1	0.99
Nonmissing	100	99.01

Because of the MISSING option on the TABLES statement, the missing values are now included in the body of the table instead of in a note on the bottom of the page. While this is useful summary information, you need to know the ID (in this case Patno) for these missing values. This brings us to the next step:

Using DATA Step Approaches to Identify and Count Missing Values

Counting missing values is not usually enough. If you have variables for which missing values are not allowed, you need to locate the observations so that the original data values can be checked and the errors corrected. A simple DATA step with PUT statements is one approach to check for any missing values for Visit, HR, Gender, and Dx. You can accomplish this easily, as shown next:

Program 7.3: Identifying Missing Values Using a DATA Step

```
title "Listing of Missing Values";
data _null_;
   file print; ***send output to the output window;
   set Clean.Patients(keep=Patno Visit HR Gender Dx);
   if missing(visit) then
      put "Missing or invalid Visit for ID " Patno;
   if missing(HR) then put "Missing or invalid HR for ID " Patno;
   if missing(Gender) then put "Missing Gender for ID " Patno;
   if missing(Dx) then put "Missing Dx for ID " Patno;
run;
```

Notice how convenient it is to use the MISSING function to test for both numeric or character missing values. Using the MISSING function also makes the program easier to read (at least for this author). Notice the message "Missing or invalid" is used for the numeric variables and the message "Missing" is used for character variables. Why? Because an invalid value for a numeric variable (i.e., a number with two decimal points or character data in a numeric field) is not really missing data—it is invalid data. However, SAS will substitute a missing value for these types of invalid data. Output from Program 7.3 is shown next:

Figure 7.4: Listing of Missing Values by Patno

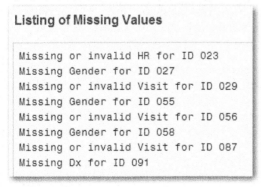

```
Listing of Missing Values

Missing or invalid HR for ID 023
Missing Gender for ID 027
Missing or invalid Visit for ID 029
Missing Gender for ID 055
Missing or invalid Visit for ID 056
Missing Gender for ID 058
Missing or invalid Visit for ID 087
Missing Dx for ID 091
```

Depending on the type of study you are conducting and whether missing values are allowed for some of these variables, you can attempt to locate the original data and make corrections if necessary.

Locating Patient Numbers for Records where Patno is Either Missing or Invalid

If you are entering your data with a data entry program such as MS/Access, you should set up rules for most or all of your variables. For the patient ID variable, a useful rule would be that it cannot be missing and it should be unique. However, what do you do if you have missing or invalid patient numbers in your data set?

Obviously, you can't list which patient number is missing because you don't have that information. One possibility is to report the patient number or numbers preceding the missing number (in the original order of data entry). If you sort the data set first, all the missing values will float to the top, and you will not have a clue as to which patients they belong to. Here is a program that prints out the two previous patient IDs when a missing ID is found:

Program 7.4: Identifying Missing or Invalid Patient Numbers

```
title "Listing of Missing or Invalid Patient Numbers";
data _null_;
   set Clean.Patients;
   ***Be sure to run this on the unsorted data set;
   file print;
   Previous_ID = lag(Patno);
   Previous2_ID = lag2(Patno);
   if missing(Patno) then
      put "Missing patient ID. Two previous ID's are:"
      Previous2_ID "and " Previous_ID /
      @5 "Missing record is number " _n_;
   else if notdigit(trimn(Patno)) then
      put "Invalid patient ID:" patno +(-1)
      ". Two previous ID's are:"
      Previous2_ID "and " Previous_ID /
      @5 "Invalid record is number " _n_;
run;
```

Although there are several solutions to listing the patient numbers from the preceding observations, the LAG function serves the purpose here. The LAG function returns the value of its argument the last time the function executed. If you execute this function for every iteration of the DATA step, it returns the value of Patno from the previous observation. The LAG2 function, when used in the same manner, returns the value of Patno from the observation before that. Remember to execute the LAG and LAG2 functions for every observation.

When Program 7.4 is run and a missing patient number is encountered, the two lagged variables will be the IDs from the previous two observations. The assumption in this program is that there are no more than three missing patient numbers in a row. If that is a possibility, you could list more than two previous patient IDs or include patient IDs following the missing one as well. Notice that we added the observation number to the output by printing the value of the internal SAS variable _N_. This provides one additional clue in finding the missing patient number.

The last part of the program uses the NOTDIGIT function to test for any invalid, non-missing values for the patient number. NOTDIGIT returns the first character in a character value that is not a digit. The TRIMN function ensures that any trailing blanks (which are treated as non-digits) are removed before the test is made. Here is the output:

Figure 7.5: Output from Program 7.4

```
Listing of Missing or Invalid Patient Numbers

Invalid patient ID:XX5. Two previous ID's are:018 and 019
    Invalid record is number 21
Missing patient ID. Two previous ID's are:028 and 029
    Missing record is number 31
```

Another approach is to list the values of all the variables for any missing or invalid patient ID. This might offer a clue about the identity of the missing ID. Using PROC PRINT with a WHERE statement makes this an easy task, as demonstrated by the SAS code in Program 7.5:

Program 7.5: Using PROC PRINT to Identify Missing or Invalid Patient Numbers

```
title "Data listing for patients with Missing or Invalid ID's";
proc print data=Clean.Patients;
   where missing(Patno) or notdigit(trimn(Patno));
run;
```

Here is the corresponding output:

Figure 7.6: Using PROC PRINT to Identify Missing or Invalid Patient Numbers

Data listing for patients with Missing or Invalid ID's

Obs	Patno	Account_No	Gender	Visit	HR	SBP	DBP	Dx	AE
21	XX5	MA93350	F	11/04/2010	69	122	190	052.040	0
31		DE56405	M	06/15/2010	87	128	98	195.920	0

Seeing the values of all the variables for these two patients might be useful in determining the correct patient number for these two individuals. Before leaving this section on DATA step detection of missing values, let's modify Program 7.3, which listed missing dates, heart rates, and adverse events, to count the number of missing values for each variable as well.

Program 7.6: Listing Missing Values and Summary of Frequencies

```
title "Listing of Missing Values and Summary of Frequencies";
data _null_;
   set Clean.Patients(keep= Patno Visit HR Gender Dx) end=Last;
   file print; ***Send output to the output window;
   if missing(Visit) then do;
      put "Missing or invalid visit date for ID " Patno;
      N_visit + 1;
   end;
   if missing(HR) then do;
      put "Missing or invalid HR for ID " Patno;
      N_HR + 1;
   end;
```

```
      if missing(Gender) then do;
         put "Missing Gender for ID " Patno;
         N_Gender + 1;
      end;

      if missing(Dx) then do;
         put "Missing Dx for ID " Patno;
         N_Dx + 1;
      end;

   if Last then
              put // "Summary of missing values" /
              25*'-' /
              "Number of missing dates = " N_Visit /
              "Number of missing HR's = " N_HR /
              "Number of missing genders = " N_Gender /
              "Number of missing Dx = " N_Dx;
   run;
```

Each time a missing value is located, the respective missing counter is incremented by 1. Because you only want to see the totals once after all the data lines have been read, use the END= option in the SET statement to create the logical variable Last. Last will be true when the last observation is being read from the Patients data set. So, along with the earlier listing, you have the additional lines of output shown next:

Figure 7.7: Output from Program 7.6

```
Listing of Missing Values and Summary of Frequencies

Missing or invalid HR for ID 023
Missing Gender for ID 027
Missing or invalid visit date for ID 029
Missing Gender for ID 055
Missing or invalid visit date for ID 056
Missing Gender for ID 058
Missing or invalid visit date for ID 087
Missing Dx for ID 091


Summary of missing values
- - - - - - - - - - - - - - - - - - - - - - - -
Number of missing dates = 3
Number of missing HR's = 1
Number of missing genders = 3
Number of missing Dx = 1
```

Searching for a Specific Numeric Value

Specific values such as 999 or 9999 are sometimes used to denote missing values. In order to demonstrate how to locate these special values, run Program 7. 7 (below) to create a data set (called Test) that contains some numeric and character variables. A number of the numeric variables have values of 999. This practice is quite popular with people who use SPSS (Statistical Package for the Social Sciences) where a statement "Assign Missing 999" automatically replaces all numeric values of 999 to a missing value. If you are given a data set that contains special values (such as 999) to represent missing values, you might want to investigate which variables have this value and how many times these values appear.

Here is the program:

Program 7.7: Creating a Test Data Set

```
***Create test data set;
data Test;
   input X Y A $ X1-X3 Z $;
datalines;
1 2 X 3 4 5 Y
2 999 Y 999 1 999 J
999 999 R 999 999 999 X
1 2 3 4 5 6 7
;
```

Below is a listing of data set Test:

Figure 7.8: Listing of Data Set Test

Listing of Data Set Test

X	Y	A	X1	X2	X3	Z
1	2	X	3	4	5	Y
2	999	Y	999	1	999	J
999	999	R	999	999	999	X
1	2	3	4	5	6	7

Note: It is a simple matter to convert the values of 999 to a SAS missing value, as you will see later.

Program 7.8 searches a SAS data set for all numeric variables set to a specific value and produces a report, which shows the variable name and the observation where the specific value was found.

The trick in this program is the VNAME function (also used in Chapter 4, Program 4.17). This function returns the variable name corresponding to an array element. Program 7.8 searches a SAS data set for a specific value. In the next section, the program is turned into a macro, making it more flexible Here is the first program:

Program 7.8: Detecting and Counting Special Values (999 in this Example)

```
***Program to detect the specified values;
title "Looking for Values of 999 in Data Set Test";
data _null_;
   set Test;
   file print;
   array Nums[*] _numeric_;
   length Varname $ 32;
   do iii = 1 to dim(Nums);
      if Nums[iii] = 999 then do;
         Varname = vname(Nums[iii]);
         put "Value of 999 found for variable " Varname
             "in observation " _n_;
      end;
   end;
   drop iii;
run;
```

One key to this program is the use of _NUMERIC_ in the ARRAY statement. Because this ARRAY statement follows the SET statement, the array Nums will contain all the numeric variables in data set Test. The next step is to examine each of the elements in the Nums array, determine if a value of 999 is found, and then determine the variable name associated with that array element. The DO loop uses the index variable iii in the hopes that there will not be any variables in the data set to be tested with that name. If there were, the DO loop counter would replace the value in the data set being tested.

Now for the trick! As you search for values of 999 for each of the numeric variables, you can use the VNAME function to return the variable name that corresponds to the array element. In this program, the variable name is stored in the variable Varname. All that is left to do is to write out the variable names and observation numbers. Output from Program 7.8 is shown below:

Figure 7.9: Output from Program 7.8

```
Looking for Values of 999 in Data Set Test

Value of 999 found for variable Y in observation 2
Value of 999 found for variable X1 in observation 2
Value of 999 found for variable X3 in observation 2
Value of 999 found for variable X in observation 3
Value of 999 found for variable Y in observation 3
Value of 999 found for variable X1 in observation 3
Value of 999 found for variable X2 in observation 3
Value of 999 found for variable X3 in observation 3
```

Creating a Macro to Search for Specific Numeric Values

Before we leave this chapter, let's modify the program above so that it only produces a summary report on variables with specific missing values. In addition, we will turn the program into a macro so that it will be more generally useful. Here is the macro:

Program 7.9: Turning Program 7.8 into a Macro

```
*Macro name: Find_Value.sas
Purpose: Identifies any specified value for all numeric variables
Calling arguments: dsn=   sas data set name
                   value= numeric value to search for
Example:  To find variable values of 999 in data set Test, use
          %Find_Value(dsn=Test, Value=999);
%macro Find_Value(Dsn=,  /* The data set name */
                  Value= /* Value to look for */ );
   title "Variables with &Value as Missing Values in Data Set &Dsn";
   data Tmp;
      set &Dsn;
      file print;
      length Varname $ 32;
      array Nums[*] _numeric_;
      do iii = 1 to dim(Nums);
         if Nums[iii] = &Value then do;
         Varname = vname(Nums[iii]);
         output;
         end;
      end;
      keep Varname;
   run;

   proc freq data=Tmp;
      tables Varname / out=Summary(keep=Varname Count)
                      nocum;
   run;

   proc datasets library=Work nolist;
      delete Tmp;
   run;
   quit;
%mend Find_Value;
```

There are several points in this macro that need explanation: First, the two calling arguments in this macro are the data set name and the value that you want to search for. The program creates an array of all numeric variables in the data set to be tested. A DO loop then tests to see if any of the numeric variables contain the specified value (such as 999). If so, the VNAME function determines the variable name corresponding to the array element and an observation is output to data set Tmp. This data set contains the single variable Varname. To help make this clear, here is a listing of data Tmp when the macro is run on data set Test looking for a value of 999 (PROC DATASETS was commented out so that we could show you data set Tmp):

Figure 7.10: Listing of Data Set Tmp

Listing of Data Set Tmp
Varname
Y
X1
X3
X
Y
X1
X2
X3

You can now use PROC FREQ to list each of the variable names for which the specific value is found, along with the frequency. The final section of the macro deletes the temporary data set Tmp. Here is the output from calling the %Find_Value macro, as follows:

```
%Find_Value(dsn=Test, Value=999)
```

Figure 7.11: Result of Running the %Find_Value Macro on Data Set Test with a Value of 999

Variables with 999 as Missing Values in Data Set Test		
Varname	Frequency	Percent
X	1	12.50
X1	2	25.00
X2	1	12.50
X3	2	25.00
Y	2	25.00

Converting Values Such as 999 to a SAS Missing Value

Although SAS does not have a statement such as "Assign Missing 999," it is quite easy to convert any special value to a SAS missing value for all numeric variables in a data set. As an example, the program below converts all values of 999 in data set Test (used in the previous section) to a SAS missing value:

Program 7.10: Converting All Values of 999 to a SAS Missing Value

```
data Set_999_to_Missing;
   set Test;
   array Nums[*] _numeric_;
   do iii = 1 to dim(Nums);
      if Nums[iii] = 999 then Nums[iii] = .;
   end;
   drop iii;
run;
```

This program is very similar to Program 7.8, except this time when a value of 999 is found, a missing value is substituted. A listing of data set Set_999_to_Missing is listed below:

Figure 7.12: Listing of Data Set Set_999_to_Missing

Listing of Data Set Set_999_to_Missing

X	Y	A	X1	X2	X3	Z
1	2	X	3	4	5	Y
2	.	Y	.	1	.	J
.	.	R	.	.	.	X
1	2	3	4	5	6	7

All the values of 999 are now SAS missing values.

Conclusions

Most SAS data sets contain missing values. These values might represent actual missing data or they might be the result of data errors (such as invalid dates). You also saw ways to search for specific values such as 999 or 888 that were purposely entered into a data file because they represented specific situations—usually missing data of some kind. Finally, you saw an easy way to convert a specific numeric value to a SAS missing value for all numeric variables in a SAS data set.

Chapter 8: Working with SAS Dates

Introduction

SAS dates seem mysterious to many SAS programmers, but by understanding how SAS dates are stored, you will see that they are really quite simple. SAS dates are stored as numeric variables and represent the number of days from a fixed point in time, January 1, 1960. The confusion develops in the many ways that SAS software can read and write dates. Typically, dates are read as *MM*/*DD*/*YYYY* or some similar form. There are informats to read almost any conceivable date notation. Regardless of how a date is read, the informat performs the conversion to a SAS date, and it is stored just like any other numeric value. If you print out a date value without a SAS date format, it will appear as a number (the number of days from January 1, 1960) rather than as a date.

When date information is not in a standard form or when you have month, day, and year values in separate variables, you can use the MDY (month-day-year) function to create a SAS date.

Before we discuss data cleaning operations involving dates, let's take a moment to talk about ways to save disk space (and speed up processing) when working with dates.

Changing the Storage Length for SAS Dates

One of the golden rules of SAS programming is to never change the storage length for SAS numeric variables from the default value of eight bytes. It's time to break that rule. If you are storing integers and you know the size of the largest integer you will be storing, you can change the storage length to fewer than eight bytes (the minimum on all systems except mainframes is three bytes).

SAS dates are always integers, and current dates are close to 21,000 days from January 1, 1960. SAS can store integers over 2,000,000, exactly, using four bytes. You use a LENGTH statement to change the length of character or numeric variables. So, if the name of your date variable is Visit, use the following SAS LENGTH statement to set the storage length of Visit to four bytes:

```
length Visit 4;
```

Caution: Do not change the storage length for any numeric variable that is not an **integer**. Lots of things can go wrong if you do.

Let's look at some ways to perform data cleaning and validation with dates.

Checking Ranges for Dates (Using a DATA Step)

Suppose you want to determine if the visit dates in the Patients data set are between January 1, 2010, and April 15, 2017. You can use a DATA step approach, much the same way you did when checking that numeric variables were within a specified range. The only difference is that the range boundaries have to be SAS dates. Let's see how this works. Program 8.1 checks for dates in the specified range and ignores missing values.

Program 8.1: Checking that a Date Is within a Specified Interval (DATA Step Approach)

```
title "Dates Before January 1, 2010 or After April 15, 2017";
data _null_;
   file print;
   set Clean.Patients(keep=Visit Patno);
   if Visit lt '01Jan2010'd and not missing(Visit) or
      Visit gt '15Apr2017'd then put Patno= Visit= mmddyy10.;
run;
```

The key to this program is the use of the date constants in the IF statement. If you want SAS to turn a date that you recognize into a SAS date (the number of days from January 1, 1960), the dates must be written in this fashion. Date constants are written as a one- or two-digit day of the month, a three-character month abbreviation, and a two- or four-digit year (always use a four-digit year), placed in single or double quotes, and followed by a lower- or uppercase 'D'. You also need to add a date format in the PUT statement so that the date will be printed in a standard date format. Output from Program 8.1 is shown next:

Figure 8.1: Output from Program 8.1

Dates Before January 1, 2010 or After April 15, 2017

```
Patno=009 Visit=03/15/1956
Patno=016 Visit=10/21/2020
```

You will need to check the visit dates for these two patients because both dates are outside the valid range of dates.

Checking Ranges for Dates (Using PROC PRINT)

You can accomplish the same objective by using a PROC PRINT statement in combination with a WHERE statement. Besides being easier to code, this approach allows you to use the keyword BETWEEN with the WHERE statement, making the logic somewhat simpler to follow. The SAS code is shown in Program 8.2.

Program 8.2: Checking that a Date Is within a Specified Interval (Using PROC PRINT)

```
title "Dates Before January 1, 2010 or After April 15, 2017";
proc print data=Clean.Patients(keep=Patno Visit) noobs;
   where Visit not between '01Jan2010'd and '15Apr2017'd and
   Visit is not missing;
   format Visit date9.;
run;
```

Output from this procedure contains the identical information as the previous DATA step approach. For variety, we chose the DATE9. date format to replace the *mmddyy*10. format that was associated with the Visit variable in the Data set.

Figure 8.2: Output from Program 8.2

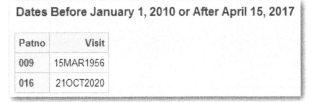

Dates Before January 1, 2010 or After April 15, 2017

Patno	Visit
009	15MAR1956
016	21OCT2020

Checking for Invalid Dates

Some of the dates in the Patients data set are missing and some are invalid dates that were converted to missing values during the input process. If you want to distinguish between the two, you must work from the raw data, not the SAS data set. If you attempt to read an invalid date with a SAS date informat, an error message will appear in the SAS Log. This is one clue that you have errors in your date values. Program 8.3 reads the raw data file Patients.txt. The resulting SAS Log follows:

Program 8.3: Reading Visit Dates from the Raw Data File

```
data Dates;
   infile "c:\Books\Clean3\Patients.txt";
   input @12 Visit mmddyy10.;
   format Visit mmddyy10.;
run;
```

The SAS Log that results from running this program is shown next:

Figure 8.3: SAS Log from Running Program 8.3

```
54    data Dates;
55        infile "c:\Books\Clean3\Patients.txt";
56        input @12 Visit mmddyy10.;
57        format Visit mmddyy10.;
58    run;

NOTE: The infile "c:\Books\Clean3\Patients.txt" is:
      Filename=c:\Books\Clean3\Patients.txt,
      RECFM=V,LRECL=32767,File Size (bytes)=4040,
      Last Modified=07Oct2016:13:24:31,
      Create Time=16Sep2016:09:54:42

NOTE: Invalid data for Visit in line 30 12-21.
RULE:       ----+----1----+----2----+----3----+----4----+----5----+-
30          029NY29028F99/99/9999 79148 88844.7900 38
Visit=. _ERROR_=1 _N_=30
NOTE: Invalid data for Visit in line 57 12-21.
57          056PA88813F10/35/2015 70108 98446.3000 38
Visit=. _ERROR_=1 _N_=57
NOTE: 101 records were read from the infile
      "c:\Books\Clean3\Patients.txt".
      The minimum record length was 38.
      The maximum record length was 38.
```

There are two dates that caused errors. The first, 99/99/9999, is a value that is sometimes used to indicate that there is a missing date—not really a good idea. The other date triggered an error message because no month has 35 days. Remember that once the number of errors exceeds a default number, errors will no longer be reported in the SAS Log. This number can be adjusted by setting the system option ERRORS=. If you have no missing date values in your raw data, any missing date value in your SAS data set must have been generated by an invalid date. You can use this idea to list missing and invalid dates. The plan is to read the date value twice; once with a SAS date informat, the next as a character value. This way, you can see the original date value that caused the error. Program 8.4 demonstrates this.

Program 8.4: Listing Missing and Invalid Dates

```
title "Listing of Missing and Invalid Dates";
data _null_;
   file print;
   infile "c:\Books\Clean3\Patients.txt";
   input @1  Patno $3.
         @12 Visit mmddyy10.
         @12 Char_Date $char10.;
   if missing(Visit) then put Patno= Char_date=;
run;
```

Here you read the date twice, first with the SAS date informat *MMDDYY*10. and then with the character informat $CHAR10. This way, even though SAS substitutes a missing value for Visit, the variable Char_Date will contain the actual characters that were entered in the date field.

This is a good place to discuss the difference between the $n and $CHAR$n$ informats: The more common $n informat will left-justify a character value. That is, if there are leading blanks in the columns being read, the resulting character value in your data set will start with the first non-blank character. The $CHAR$n$ informat reads the specified columns as is and leaves leading blanks, if there are any, in the character value. We chose to use the $CHAR$n$ informat in the program above so that you could see exactly what was contained in the 10 columns holding the date.

Running Program 8.4 results in the following output:

Figure 8.4: Output from Program 8.4

```
Listing of Missing and Invalid Dates

Patno=029 Char_Date=99/99/9999
Patno=056 Char_Date=10/35/2015
Patno=087 Char_Date=
```

If you want to ignore missing values in the raw data and only identify invalid dates, you only need to make a slight change, as shown in Program 8.5:

Program 8.5: Listing Invalid Dates

```
title "Listing of iInvalid Dates";
data _null_;
    file print;
    infile "c:\Books\Clean3\patients.txt";
    input @1  Patno $3.
          @12 Visit mmddyy10.
          @12 Char_Date $char10.;
    if missing(Visit) and not missing(Char_Date) then
        put Patno= Char_Date=;
run;
```

If the numeric date value is missing and the character value holding the date is not missing, then you have located a non-missing, invalid value in the raw data, and the dates will be printed. Output from this program is shown below:

Figure 8.5: Output from Program 8.5

```
Listing of iInvalid Dates

Patno=029 Char_Date=99/99/9999
Patno=056 Char_Date=10/35/2015
```

Notice that the observation with a missing visit date is not listed.

Working with Dates in Nonstandard Form

Although SAS can read dates in almost every conceivable form, there may be times when you have date information for which there is no SAS informat or you are given a SAS data set where, instead of a variable representing a date, you have separate variables for month, day, and year. In either case, there is an easy solution—use the MDY (month-day-year) function.

As an example, suppose you have a month value (a number from 1 to 12) in columns 6-7, a day of the month value (a number from 1 to 31) in columns 13-14, and a four-digit year value in columns 20-23. Enter the three variable names for the month, day, and year as arguments to the MDY function, and it will return a SAS date. Program 8.6 demonstrates how this works:

Program 8.6: Demonstrating the MDY Function to Read Dates in Nonstandard Form

```
data Nonstandard;
   input Patno $ 1-3 Month 6-7 Day 13-14 Year 20-23;
   Date = mdy(Month,Day,Year);
   format date mmddyy10.;
datalines;
001  05      23       1998
006  11      01       1998
123  14      03       1998
137  10       .       1946
;
title "Listing of data set Nonstandard";
proc print data=Nonstandard;
   id Patno;
run;
```

Notice that an invalid month value (observation three) and a missing day value (observation four) were included intentionally. The listing of the data set Nonstandard follows:

Figure 8.6: Listing of Data Set Nonstandard

Listing of data set Nonstandard

Patno	Month	Day	Year	Date
001	5	23	1998	05/23/1998
006	11	1	1998	11/01/1998
123	14	3	1998	.
137	10	.	1946	.

In the two cases where a date could not be computed, a missing value was generated. Inspection of the SAS Log also shows that the MDY function had an invalid value and a missing value.

Creating a SAS Date When the Day of the Month Is Missing

Some of your date values might be missing the day of the month, but you would still like to create a SAS date by using either the 1st or the 15th of the month as the day. There are two possibilities here. One method is to use the *MONYY* informat that reads dates in the form of a three-character month name and a two- or four-digit year. If your dates are in this form, SAS will create a SAS date using the first of the month as the day value. The other method of creating a SAS date from only month and year values is to use the MDY function, substituting a value such as 15, for the day argument. An example is shown in Program 8.7:

Program 8.7: Creating a SAS Date When the Day of the Month Is Missing

```
data No_Day;
   input @1  Date1 monyy7.
         @8  Month 2.
         @10 Year 4.;
   Date2 = mdy(Month,15,Year);
   format Date1 Date2 mmddyy10.;
datalines;
JAN98  011998
OCT1998101998
;
title "Listing of data set No_Day";
proc print data=No_Day;
run;
```

Date1 is a SAS date created by the *MONYY* SAS informat; Date2 is created by the MDY function, using the 15th of the month as the missing day value. Output from PROC PRINT is shown next:

Figure 8.7: Output from Program 8.7

Listing of data set No_Day

Obs	Date1	Month	Year	Date2
1	01/01/1998	1	1998	01/15/1998
2	10/01/1998	10	1998	10/15/1998

Let's extend this idea a bit further. Suppose that most of your dates have month, day, and year values but, for any date where the only piece missing is the day of the month, you want to substitute the 15th of the month. Program 8.8 will accomplish this goal:

Program 8.8: Substituting the 15th of the Month When the Day of the Month Is Missing

```
data Miss_Day;
   input @1  Patno  $3.
         @4  Month  2.
         @6  Day    2.
         @8  Year   4.;
   if not missing(Day) then Date = mdy(Month,Day,Year);
   else Date = mdy(Month,15,Year);
   format Date mmddyy10.;
```

```
datalines;
00110211998
00205  1998
00344  1998
;

title "Listing of data set Miss_Day";
proc print data=Miss_Day;
run;
```

If the day value is not missing, the MDY function uses all three values of month, day, and year to compute a SAS date. If the day value is missing, the 15th of the month is used. As before, if there is an invalid date (as there is for patient 003), a missing date value is generated. Here are the three observations created by this program:

Figure 8.8: Output from Program 8.8

Listing of data set Miss_Day

Obs	Patno	Month	Day	Year	Date
1	001	10	21	1998	10/21/1998
2	002	5	.	1998	05/15/1998
3	003	44	.	1998	.

Hey, want to see something clever? My SAS friend Mark Jordan (AKA SAS Jedi) came up with this clever alternative to Program 8.8:

Program 8.9: Clever Alternative to Program 8.8

```
data Miss_Day;
   input @1  Patno  $3.
         @4  Month  2.
         @6  Day    2.
         @8  Year   4.;
   Date = mdy(Month,coalesce(Day,15),Year);
   format Date mmddyy10.;
datalines;
00110211998
00205  1998
00344  1998
;
```

The COALESCE function can take any number of arguments. and it returns the value of the first non-missing value in the list of arguments. So, if Day is not missing, the function returns its value—if Day is missing, it returns a 15. Think of using the COALESCE function when you are testing for a missing value and performing two actions—one action if the test for a missing value is true and another action if the test for a missing value is false. Output from this program is identical to the listing shown in Figure 8.8.

Suspending Error Checking for Known Invalid Dates

As you saw earlier, invalid date values can fill your SAS Log with lots of errors. There are times when you know that invalid date values were used to represent missing dates or other specific values. If you would like to prevent the automatic listing of date errors in the SAS Log, you can use the double question mark (??) modifier in your INPUT statement or with the INPUT function. You place the two question marks between the variable name and the informat. This modifier prevents the NOTES and data listings to be printed in the SAS Log and also keeps the SAS internal variable _ERROR_ at 0.

Program 8.10 uses the ?? modifier in the INPUT statement to prevent error messages from printing in the SAS Log.

Program 8.10: Suspending Error Checking for Known Invalid Dates by Using the ?? Informat Modifier

```
data Dates;
    infile "c:\Books\Clean3\Patients.txt";
    input @12 Visit ?? mmddyy10.;
    format Visit mmddyy10.;
run;
```

When this program is run, there will be no error messages in the SAS Log caused by invalid dates.

> Only turn off SAS error checking when you plan to detect errors in other ways or when you already know all about your invalid dates.

Conclusions

Hopefully, after reading this chapter, you will feel more comfortable working with SAS dates. If you have more complicated programming tasks that involve date and/or time functions, I highly recommend Derek Morgan's book, *The Essential Guide to SAS Dates and Times*, 2nd *Edition*, available from SAS Press.

Chapter 9: Looking for Duplicates and Checking Data with Multiple Observations per Subject

Introduction

There are a number of tasks that are unique to data sets where you have multiple observations per subject. Luckily, SAS has all the programming tools you need to work with these types of data.

Besides checking for invalid data values in a data set, it might be necessary to check for either duplicate IDs or duplicate observations. *Duplicate observations* are observations where the values of all the variables are identical. This is an easy problem to fix—just eliminate the duplicates (although you might want to find out how the duplicates got there). Duplicate IDs with different data values present another problem. One possible cause is that the same ID was used for more than one person. Another possibility is that different data values were entered more than once for the same person. A third possibility is that multiple records per ID are expected. There are several ways to detect and eliminate unwanted duplicates in a SAS data set. This chapter explores some of them. (For more detailed information about working with data sets with multiple observations per subject, please take a look at "*Learning SAS by Example: A Programmer's Guide*, by this author.)

Eliminating Duplicates by Using PROC SORT

Imagine that you have a data set where each patient is supposed to be represented by a single observation. To demonstrate what happens when you have multiple observations with the same ID, some duplicates in the Patients data set were included on purpose. These observations are shown in the next figure:

Figure 9.1: Listing of Duplicates in Data Set Duplicates

Listing of Duplicates in the Patients Data Set

Patno	Account_No	Gender	Visit	HR	SBP	DBP	Dx	AE
005	DE00080	F	04/08/2012	91	106	84	078.160	0
005	DE00080	F	04/08/2012	91	106	84	078.160	0
007	NJ90043	M	08/06/2010	83	130	102	564.870	0
007	VT56383	F	01/13/2014	63	128	80	640.260	1
050	NJ87682	M	11/01/2010	43	112	62	685.680	0
050	PA37838	M	03/09/2011	32	118	90	692.470	0

Notice that patient number 005 has a true duplicate observation. For patient numbers 007 and 050, the duplicate IDs contain different values.

Two very useful options of PROC SORT are NODUPKEY and NODUPRECS (also called NODUP). The NODUPKEY option automatically eliminates multiple observations where the BY variables have the same value. For example, to automatically eliminate multiple patient IDs (Patno) in the Patients data set (which you probably would not want to do—this is for illustration only), you could use PROC SORT with the NODUPKEY option as shown in Program 9.1.

Program 9.1: Demonstrating the NODUPKEY Option of PROC SORT

```
proc sort data=Clean.Patients out=Single nodupkey;
   by Patno;
run;

title "Data Set Single - Duplicated ID's Removed from Patients";
proc print data=Single;
   id Patno;
run;
```

Notice that two options, OUT= and NODUPKEY, are used here. The OUT= option is used to create the new data set Single, leaving the original data set Patients unchanged. This is considered good programming practice. Shown next is a partial listing of the Single data set.

Figure 9.2: Partial Listing of Data Set Single

Data Set Single - Duplicated ID's Removed from Patients

Patno	Account_No	Gender	Visit	HR	SBP	DBP	Dx	AE
	DE56405	M	06/15/2010	87	128	98	195.920	0
001	CT14882	M	06/12/2012	69	124	86	713.410	0
002	MD78461	M	06/04/2010	76	130	80	047.570	1
003	DE51381	f	06/22/2013	70	56	70	108.510	0
004	CT37146	M	05/18/2013	76	112	84	669.860	0
005	DE00080	F	04/08/2012	91	106	84	078.160	0
006	DE37709	M	07/27/2014	71	104	88	967.570	0
007	NJ90043	M	08/06/2010	83	130	102	564.870	0
008	PA67069	F	09/28/2013	79	124	72	020.120	0
	MD68313				132	64		

Notice that patients 005 and 007 now have a single observation. For patient 007, it is the first of the two duplicates show in Figure 9.1. (Patient 050, not shown in this figure, also has a single observation.)

The NODUPKEY option eliminated the second observation for each of the three duplicate IDs. The only indication that duplicates were removed is in the NOTE in the SAS Log, which is shown next:

Figure 9.3: Listing of the SAS Log After Running Program 9.1

```
35    proc sort data=Clean.Patients out=Single nodupkey;
36        by Patno;
37    run;

NOTE: There were 101 observations read from the data set CLEAN.PATIENTS.
NOTE: 3 observations with duplicate key values were deleted.
NOTE: The data set WORK.SINGLE has 98 observations and 9 variables.
NOTE: PROCEDURE SORT used (Total process time):
      real time            0.00 seconds
      cpu time             0.00 seconds
```

This method of looking for duplicate IDs is really only useful if the SAS Log shows that no duplicates were removed. If the SAS Log shows duplicate key values were deleted and duplicates were not expected, you need to see which IDs had duplicate data and the nature of that data.

If you use the NODUPKEY option with more than one BY variable, only those observations with identical values on each of the BY variables will be deleted. For example, if you sort by patient number (Patno) and visit date (Visit), only the duplicate observation for patient number 005 will be deleted when you use the NODUPKEY option. That is because the two observations for patient number 005 are the only ones with the same patient number and visit date.

The option NODUPRECS (stands for no duplicate records) also deletes duplicates but only for observations where all the variables have identical values. Program 9.2 demonstrates this option.

Program 9.2: Demonstrating the NODUPRECS Option of PROC SORT

```
proc sort data=Clean.Patients out=Single noduprecs;
   by Patno;
run;
```

Listing the data set SINGLE that is created by this procedure shows that the second observation for patient number 005 was deleted. (See the partial listing below:)

Figure 9.4: Output from Program 9.2

Demonstrating the NODUPRECS Option of PROC SORT

Patno	Account_No	Gender	Visit	HR	SBP	DBP	Dx	AE
	DE56405	M	06/15/2010	87	128	98	195.920	0
001	CT14882	M	06/12/2012	69	124	86	713.410	0
002	MD78461	M	06/04/2010	76	130	80	047.570	1
003	DE51381	f	06/22/2013	70	56	70	108.510	0
004	CT37146	M	05/18/2013	76	112	84	669.860	0
005	DE00080	F	04/08/2012	91	106	84	078.160	0
006	DE37709	M	07/27/2014	71	104	88	967.570	0
007	NJ90043	M	08/06/2010	83	130	102	564.870	0
007	VT56383	F	01/13/2014	63	128	80	640.260	1
008	PA67069	F	09/28/2013	79	124	72	020.120	0
009	MD68313	F	03/15/1956	82	132	64	894.400	0

Even though the NODUPRECS option worked as expected in this example, there is a slight problem with this option: It doesn't always work the way you want or expect. An example of this strange behavior is shown next, followed by an explanation and a way to avoid this problem.

Demonstrating a Possible Problem with the NODUPRECS Option

The NODUPRECS option, as defined in SAS documentation, removes **successive** duplicates from a SAS data set. A small data set called Multiple, listed below, can be used to demonstrate that the NODUPRECS option may not always remove duplicate observations (records) from a SAS data set. First, take a look at the listing of data set Multiple in Figure below:

Figure 9.5: Listing of Data Set Multiple

Listing of Data Set Multiple

ID	X	Y
001	1	2
006	1	2
009	1	2
001	3	4
001	1	2
009	1	2
001	1	2

The four observations for ID 001 are highlighted. Now, run PROC SORT with the NODUPRECS option:

Program 9.3: Running PROC SORT with the NODUPRECS Option

```
proc sort data=Multiple out=Strange noduprecs;
   by ID;
run;
```

Here is a listing of the sorted data set (Strange):

Figure 9.6: Listing of Data Set Strange

Listing of Data Set Strange

ID	X	Y
001	1	2
001	3	4
001	1	2
006	1	2
009	1	2

The duplicate record for ID 009 was removed. However, notice that there are still two duplicate observations in the output data set (ID 001 with X=1 and Y=2). To see what happened and why, let's first see how the sort process works. To do that, let's run PROC SORT on data set Multiple, without the NODUPRECS option. Here is the result:

Figure 9.7: Result of Running PROC SORT without the NODUPRECS Option

Listing of Data Set Multiple_Sorted

ID	X	Y
001	1	2
001	3	4
001	1	2
001	1	2
006	1	2
009	1	2
009	1	2

Only the highlighted observations (ID=001, X=1, Y=2) are **successive** duplicates. By definition, the NODUPRECS option removes one of these two observations, leaving two observations (the first and third) with identical data. How do you fix this problem?

To be sure that you remove all duplicates with the NODUPRECS option, you might want to include several BY variables so that you can feel confident that any duplicate records are successive. You could use _ALL_ for your list of BY variables, guaranteeing that all duplicate observations are successive. However, with a large number of variables and observations, this might require extensive computer resources. Using several BY variables, especially ones with many unique values, should usually be enough.

For example, if you sort data set Multiple by ID and X and use the NODUPRECS option, all duplicate records for ID 001 will be removed.

My thanks to Mike Zdeb, who first brought the unexpected behavior of the NODUPRECS option to my attention.

Reviewing First. and Last. Variables

When you have one or more observations per subject (or patient or any other BY group), you need some special programming tools—and SAS provides them. As an example, suppose that you have a SAS data set consisting of patient visits to a clinic. Data set Clinic_Visits (shown below) is such a data set.

So that you can create this data set and play with it, the program to generate the data set is included along with the listing:

Program 9.4: Creating and Listing Data Set Clinic_Visits

```
data Clean.Clinic_Visits;
   informat ID $3. Date mmddyy10.;
   input ID Date HR SBP DBP;
   format Date date9.;
datalines;
001 11/11/2016 80 120 76
001 12/24/2016 78 122 78
```

```
002 1/3/2017 66 140 88
003 2/2/2017 80 144 94
003 3/2/2017 78 140 90
003 4/2/2017 78 134 78
004 11/15/2016 66 118 78
004 11/15/2016 64 116 76
005 1/5/2017 72 132 82
005 3/15/2017 74 134 84
;
title Listing of Data Set Clinic_Visits;
proc print data=Clean.Clinic_Visits;
    id ID;
run;
```

Here is a listing of data set Clinic_Visits:

Figure 9.8: Listing of Data Set Clinic_Visits

Listing of Data Set Clinic_Visits

ID	Date	HR	SBP	DBP
001	11NOV2016	80	120	76
001	24DEC2016	78	122	78
002	03JAN2017	66	140	88
003	02FEB2017	80	144	94
003	02MAR2017	78	140	90
003	02APR2017	78	134	78
004	15NOV2016	66	118	78
004	15NOV2016	64	116	76
005	05JAN2017	72	132	82
005	15MAR2017	74	134	84

Notice that patients had from one to three visits.

It is often useful to know when you are processing the first visit or the last visit for each patient. For example, at the first visit you might want to initialize some counters—at the last visit you might want to perform an output.

SAS has the tools to do this. If you want to create First. and Last. variables for the Clinic_Visits data set, first sort the data set by ID. (Note: Even though the data set is already in Date order within each ID, you might as well sort it by ID and Date, just to be sure.) Once the data set is sorted, you proceed as follows:

Program 9.5: Creating First. and Last. Variables

```
proc sort data=Clean.Clinic_Visits;
   by ID Date;
run;

title "Examining First.ID and Last.ID";
data Clinic_Visits;
   set Clean.Clinic_Visits;
   by ID;
   file print;
   put @1 ID= @10 Date= @25 First.ID= @38 Last.ID=;
run;
```

The key is to follow the SET statement with a BY statement. It is the BY statement that creates the First. and Last. variables. For each variable listed on the BY statement, SAS creates two variables. In this program, the single BY variable is ID, so the two temporary variables created by SAS are called First.ID and Last.ID.

You can see the results of the PUT statement in Program 9.5 in the listing below. Note: You cannot use PROC PRINT to list the variables First.ID and Last.ID because they are temporary variables and exist only for the duration of the DATA step.

Figure 9.9: Output from the PUT Statement in Program 9.5

```
Examining First.ID and Last.ID


ID=001    Date=11NOV2016 FIRST.ID=1    LAST.ID=0
ID=001    Date=24DEC2016 FIRST.ID=0    LAST.ID=1
ID=002    Date=03JAN2017 FIRST.ID=1    LAST.ID=1
ID=003    Date=02FEB2017 FIRST.ID=1    LAST.ID=0
ID=003    Date=02MAR2017 FIRST.ID=0    LAST.ID=0
ID=003    Date=02APR2017 FIRST.ID=0    LAST.ID=1
ID=004    Date=15NOV2016 FIRST.ID=1    LAST.ID=0
ID=004    Date=15NOV2016 FIRST.ID=0    LAST.ID=1
ID=005    Date=05JAN2017 FIRST.ID=1    LAST.ID=0
ID=005    Date=15MAR2017 FIRST.ID=0    LAST.ID=1
```

The variable First.ID is true (equal to 1) for the first visit and false (0) otherwise—Last.ID is true for the last visit and false otherwise. Notice that the two variables First.ID and Last.ID are both true for the single observation for patient 002. Why? Because it is both the first and last visit for this patient. This is a useful way to identify patients with only a single visit.

Detecting Duplicates by Using DATA Step Approaches

Let's use the information about First. and Last. variables from the last section to explore ways to detect duplicate IDs and duplicate observations in the Patients data set. To see how this works, look at Program 9.6 that prints out all observations that have duplicate patient numbers.

Program 9.6: Identifying Duplicate IDs

```
proc sort data=Clean.Patients out=Tmp;
   by Patno;
run;

data Duplicates;
   set Tmp;
   by Patno;
   if First.Patno and Last.Patno then delete;
run;
```

You first sort the data set by the ID variable (Patno). In the above program, the original data set was left intact and a new data set (Tmp) was created for the sorted observations. After you have a sorted data set, a short DATA step will remove patients who have a single observation, leaving a data set of duplicates. In this example, there is only one BY variable (Patno), so the two temporary SAS variables First.Patno and Last.Patno are created. If First.Patno and Last.Patno are both true, there is only one observation for that patient number and that patient is removed, leaving all the observations with duplicate patient numbers. (Note: Because First.Patno and Last.Patno are logical variables with values of true or false, you do not need to write `if First.Patno=1 and Last.Patno=1`, unless you prefer that syntax.)

A listing of data set Duplicates is shown next:

Figure 9.10: Listing of Data Set Duplicates

Listing of duplicates from Data Set Clean.Patients

Patno	Account_No	Gender	Visit	HR	SBP	DBP	Dx	AE
005	DE00080	F	04/08/2012	91	106	84	078.160	0
005	DE00080	F	04/08/2012	91	106	84	078.160	0
007	VT56383	F	01/13/2014	63	128	80	640.260	1
007	NJ90043	M	08/06/2010	83	130	102	564.870	0
050	PA37838	M	03/09/2011	32	118	90	692.470	0
050	NJ87682	M	11/01/2010	43	112	62	685.680	0

Using PROC FREQ to Detect Duplicate IDs

Another way to find duplicates uses PROC FREQ to count the number of observations for each value of the patient ID variable (Patno) and create an output data set. You use the OUT= option on the TABLES statement to create a SAS data set that contains the value of the variable listed on the TABLES statement (Patno) and the frequency count for this variable. (PROC FREQ uses the variable name Count to hold the frequency information.) After you have this information, you can use it to list patients who have duplicate values of Patno. To demonstrate how this works, run Program 9.7 to create a data set that identifies observations with duplicate patient numbers:

Program 9.7: Using PROC FREQ and an Output Data Set to Identify Duplicate IDs

```
proc freq data=clean.patients noprint;
   tables Patno / out=Duplicates(keep=Patno Count
                              where=(Count gt 1));
run;
```

The NOPRINT option is used because you want PROC FREQ to create an output data set, but you do not need printed output. The KEEP= data set option says you want to keep Patno and Count. (Note: The only other variable in the output data set reports percentages, and it is called Percent.) To be efficient, you use a WHERE= data set option to restrict the observations in the output data set to those where the frequency (Count) is greater than 1 rather than writing a separate DATA step to do this.

A listing of data set Duplicates is shown next:

Figure 9.11: Listing of Data Set Duplicates (using PROC FREQ)

Listing of Patiets witb Duplicate Values

Patno	COUNT
005	2
007	2
050	2

If you want to see all the actual duplicate observations (as in Figure Figure 9.1), you can merge the Duplicates data set with the original Patient data set like this:

Program 9.8: Using the Duplicates Data Set to Select Observations from the Patients Data Set

```
proc sort data=Clean.Patients out=Tmp;
   by Patno;
run;

proc sort data=Duplicates;
   by Patno;
run;

data Duplicate_Obs;
   merge Tmp Duplicates(in=In_Duplicates drop=Count);
   by Patno;
   if In_Duplicates;
run;
```

You sort both the original data set Patients (creating the temporary data set Tmp) and the Duplicates data set by Patno. The final DATA step merges the two data sets. The key to the entire program is the IN= data set option. The Duplicates data set only contains patient numbers where the value of Count is greater than 1. The logical variable In_Duplicates, created by this IN= data set option, is true whenever the Duplicates data set is making a contribution to the merge. Using the temporary variable In_Duplicates with a subsetting IF statement selects the appropriate observations from the Patients data set. This data set is identical to the one shown in Figure Figure 9.1.

Working with Data Sets with More Than One Observation per Subject

Let's use data set Clinic_Visits (where you expect multiple visits for patients) to demonstrate tasks that are traditionally performed on data sets such as this one (this data set is displayed in figure 9.8). Some programmers refer to data sets where there are multiple observations per subject as *longitudinal data sets*.

Notice that there are from one to three observations for each patient. Also, notice that patient 004 has two observations with the same Date (and different values for HR, SBP, and DBP). To detect this situation, use the variables First. and Last., except with two BY variables instead of one. Using First. and Last. with several BY variables can be confusing. So, before we use them to identify the patient who had two visits on the same day, let's examine the value of the variables First.ID, Last.ID, First.Date, and Last.Date by running the following program:

Program 9.9: Examining First. and Last. Variables with Two BY Variables

```
proc sort data=Clean.Clinic_Visits;
   by ID Date;
run;

title "Examining First. and Last. Variables with Two BY Variables";
data Clinic_Visits;
   set Clean.Clinic_Visits;
   by ID Date;
   file print;
   put @1 ID= @8 Date= @24 First.ID= @36 Last.ID=
       @48 First.Date= @62 Last.Date=;
run;
```

You sort as before by executing a two-level sort, by ID and Date. This places the observations in the data set in ID and Date order. You then follow the SET statement with a BY statement, listing both variables (ID and Date).

Here is the listing of the First. and Last. variables for ID and Date:

Figure 9.12: First. and Last. Variables with More Than Two BY Variables

```
Examining First. and Last. Variables with Two BY Variables

ID=001 Date=11NOV2016 FIRST.ID=1 LAST.ID=0 FIRST.Date=1 LAST.Date=1
ID=001 Date=24DEC2016 FIRST.ID=0 LAST.ID=1 FIRST.Date=1 LAST.Date=1
ID=002 Date=03JAN2017 FIRST.ID=1 LAST.ID=1 FIRST.Date=1 LAST.Date=1
ID=003 Date=02FEB2017 FIRST.ID=1 LAST.ID=0 FIRST.Date=1 LAST.Date=1
ID=003 Date=02MAR2017 FIRST.ID=0 LAST.ID=0 FIRST.Date=1 LAST.Date=1
ID=003 Date=02APR2017 FIRST.ID=0 LAST.ID=1 FIRST.Date=1 LAST.Date=1
ID=004 Date=15NOV2016 FIRST.ID=1 LAST.ID=0 FIRST.Date=1 LAST.Date=0
ID=004 Date=15NOV2016 FIRST.ID=0 LAST.ID=1 FIRST.Date=0 LAST.Date=1
ID=005 Date=05JAN2017 FIRST.ID=1 LAST.ID=0 FIRST.Date=1 LAST.Date=1
ID=005 Date=15MAR2017 FIRST.ID=0 LAST.ID=1 FIRST.Date=1 LAST.Date=1
```

Examine the variables First.Date and Last.Date for ID 004. In the second observation for this ID, the variable First.Date is equal to 0—it is not the first value of 15Nov2016 for that ID.

You can use this information to identify any subject in the Clinic_Visits data set who had two visits on the same date. The program to do this is almost identical to Program 9.6 except that you are looking for observations where the SAS temporary variables First.Date and Last.Date are not both equal to 1 (see Program 9.10 below):

Program 9.10: Identifying Patients with Two Visits on the Same Date

```
proc sort data=Clean.Clinic_Visits;
   by ID Date;
run;

title "Patient with Two Visits on the Same Date";
data Duplicate_Dates;
   set Clean.Clinic_Visits;
   by ID Date;
   if First.Date and Last.Date then delete;
run;
```

Only the two identical visit dates for patient 004 will be selected for the data set Duplicate_Dates, as demonstrated in the listing shown next:

Figure 9.13: Listing of Data Set Duplicate Dates

Patient with Two Visits on the Same Date

ID	Date	HR	SBP	DBP
004	15NOV2016	66	118	78
004	15NOV2016	64	116	76

Identifying Subjects with *n* Observations Each (DATA Step Approach)

Besides identifying duplicates, you might need to verify that there are *n* observations per subject in a raw data file or in a SAS data set. For example, if each patient in a clinical trial was seen twice, you might want to verify that there are two observations for each patient ID in the data set. You can accomplish this task by using a DATA step approach or by using PROC FREQ, the same two methods used earlier to detect duplicates. First, let's look at the DATA step approach.

You can use the Clinic_Visits data set (see Figure 9.8) to demonstrate how to identify any ID that does not have exactly two visits. This is easily accomplished by using the First. and Last. temporary variables.

Take a look at the program that follows:

Program 9.11: Using a DATA Step to List All IDs for Patients Who Do Not Have Exactly Two Observations

```
proc sort data=Clean.Clinic_Visits(keep=ID) out=Tmp;
   by ID;
run;

title "Patient ID's for patients with other than two observations";
data _null_;
   file print;
   set Tmp;
   by ID; ❶
   if First.ID then n = 0; ❷
   n + 1; ❸
   if last.ID and n ne 2 then put
       "Patient number " ID "has " n "observation(s)."; ❹
run;
```

The first step is to sort the data set by patient ID because you will use a BY ID statement in the DATA_NULL_ step.

In this example, you create a new data set to hold the sorted observations (because it only contains the one variable, ID). Notice that you only need to keep the patient ID variable in the temporary data set (Tmp) because that provides sufficient information to count the number of observations per patient. If each observation contains a large number of variables, this will save processing time.

Use a DATA _NULL_ step to do the counting and output the invalid patient IDs:

❶ Follow the SET statement with a BY statement to create the variables First.ID and Last.ID.

❷ When you are processing the first observation for any patient, the temporary variable First.ID will be true and *n* is set to 0.

❸ The counter (*n*) is incremented by 1. This statement, called a *SUM statement* (notice that there is no equal sign), causes *n* to be automatically retained and initialized to 0.

❹ When you reach the last observation for any patient ID, output an error statement if your counter is not equal to 2. The next listing shows that this program works properly:

Figure 9.14: Output from Program 9.11

```
Patient ID's for patients with other than two observations

Patient number 002 has 1 observation(s).
Patient number 003 has 3 observation(s).
```

What if a patient with three observations really only had two visits but one of these visits was duplicated by mistake? You should probably run one of the programs to detect duplicates before running this program.

Identifying Subjects with *n* Observations Each (Using PROC FREQ)

You can use PROC FREQ to count the number of observations per subject, just as you did to detect duplicates. Use the variable Count to determine the number of observations for each value of ID, as shown in the next program:

Program 9.12: Using PROC FREQ to List All IDs for Patients Who Do Not Have Exactly Two Observations

```
proc freq data=Clean.Clinic_Visits noprint;
   tables ID / out=Duplicates(keep=ID Count
                              where=(Count ne 2));
run;

title "Patient ID's for Patients with Other than Two Observations";
proc print data=Duplicates noobs;
run;
```

The output data set from PROC FREQ (Duplicates) contains the variable ID and the frequency (Count), and there is one observation for each patient who did not have exactly two visits. All that is left to do is to print out the observations in data set Duplicates. Output from Program 9.12 is shown next:

Figure 9.15: Output from Program 9.12

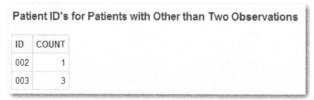

Patient ID's for Patients with Other than Two Observations

ID	COUNT
002	1
003	3

It is usually easier to let a PROC do the work, as in this example, rather than doing all the work yourself with a DATA step. But, then again, some of us really enjoy the DATA step.

Conclusions

When you have data sets with duplicate observations or with duplicate ID values, you can use DATA step programming or SAS procedures such as SORT or FREQ to detect and/or eliminate them. If you have data that contains multiple observations per subject (or some other unit), there are special programming tools such as First. and Last. variables that allow you to determine the number of observations per subject or to identify conditions (such as two patient visits on the same day) that you consider to be errors.

Chapter 10: Working with Multiple Files

Introduction

This chapter addresses data validation techniques where multiple files or data sets are involved. It is impossible to anticipate all the possible multi-file rules you might need to verify, but one task that you might need to accomplish is to ensure that an ID exists in each of several files. This chapter is devoted to that task. The last section of this chapter describes a macro that you can use to test if every ID exists in two or more SAS data sets.

Checking for an ID in Each of Two Files

One requirement of a large project might be that a particular ID exists in each of several SAS data sets. Let's start out by demonstrating how you can easily check that an ID is in each of two data sets. This will be generalized later to include an arbitrary number of data sets.

The technique demonstrated in this section is to merge the two data sets in question using the ID variable as a BY variable. The key to the program is the IN= data set option that sets a logical variable to true or false, depending on whether the data set provides values to the current observation being merged. An example will make this clear. Program 10.1 contains the SAS statements to create two SAS data sets for testing purposes.

Program 10.1: Creating Two Test Data Sets

```
data One;
   input Patno x y;
datalines;
1 69 79
2 56 .
3 66 99
5 98 87
12 13 14
;
data Two;
   input Patno z;
datalines;
```

```
1  56
3  67
4  88
5  98
13 99
;
```

Note: In the examples in this chapter, the variable Patno is a numeric variable—in the original Patients data set, Patno was a character variable. The techniques developed in this chapter work as long as the ID variable has the same name in the multiple data sets being tested and they are the same type (character or numeric). If that is not the case, you will have to use a RENAME= data set option if the ID names are not identical in all files and either a PUT or INPUT function to perform a numeric-to-character or a character-to-numeric conversion if the data types do not match.

Below are listings of the two data sets:

Figure 10.1: Data Sets One and Two

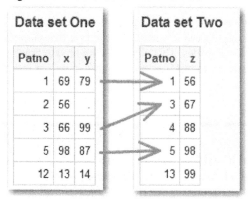

Notice that IDs 2 and 12 are in data set One but not in data set Two; IDs 4 and 13 are in data set Two but not in data set One. Program 10.2 gives detailed information on the unmatched IDs.

Program 10.2: Identifying IDs Not in Each of Two Data Sets

```
proc sort data=One;
   by Patno;
run;

proc sort data=Two;
   by Patno;
run;

title "Listing of Missing ID's";
data _null_;
   file print;
   merge one(in=In_One)
         two(in=In_Two)   end=Last;  ❶
```

```
    by Patno;  ❷

    if not In_One then do;  ❸
       put "ID " Patno "is not in data set One";
       n + 1;
    end;

    else if not In_Two then do;
       put "ID " Patno "is not in data set Two";
       n + 1;
    end;

    if Last and n eq 0 then
       put "All ID's match in both files";  ❺
run;
```

Before you can merge the two data sets, they must first be sorted by the BY variable (unless the two data sets are indexed or already sorted):

❶   The MERGE statement is the key to this program. Each of the data set names is followed by the data set option IN=*logical_variable*. In addition, the END= *variable_name* MERGE option creates a logical variable that is set to 1 (true) when the last observation from the longest data set has been processed.

❷   Using a MERGE statement would be useless in this application without the BY statement.

Because it is almost always an error to omit a BY statement following a MERGE statement, consider using the SAS system option MERGENOBY to check for this possibility. Valid values for this option are NOWARN (the default), WARN, and ERROR. If you use WARN, SAS performs the merge but adds a warning message in the SAS Log. If you use ERROR, the DATA step stops and an error message is printed to the SAS Log. It is strongly recommend that you set this option to ERROR, like this:

```
options mergenoby=error;
```

If, for some reason, you actually want to perform a merge without a BY statement, you can set this option to NOWARN before the DATA step and back to ERROR following the DATA step.

> If both data sets have multiple observations with the same ID (and the number of observations is not the same in both data sets), the merge will not work properly. If you only expect a single observation for each ID, you would be wise to check for unexpected duplicates as described in the previous chapter or to use the NODUPKEY option of PROC SORT on both of the data sets.

Let's "play computer" to see how this program works. Both data sets contain an observation for Patno=1. Therefore, in the first iteration of the DATA step, the two logical variables In_One and In_Two are both true, neither of the IF statements ❸ or ❹ is true, and a message is not printed to the output file. The next value of Patno is 2, from data set One. Because this value is not in data set Two, the value of In_One is true and the value of In_Two is false. Therefore, statement ❹ is true and the appropriate message is printed. When you reach Patno=4, which exists in data set Two but not in data set One, statement ❸ is true and its associated message is printed. Note that anytime a value of Patno is missing from one of the data sets, the variable *n* is incremented. When you reach the last observation in the longest data set being merged, the logical variable Last. is true. If there are no ID errors, *n* will still be 0 and statement ❺ will be true. When you run Program 10.2, the following output is obtained:

Figure 10.2: Output from Program 10.2

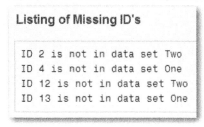

Listing of Missing ID's

```
ID 2 is not in data set Two
ID 4 is not in data set One
ID 12 is not in data set Two
ID 13 is not in data set One
```

All the ID errors are correctly displayed. If you want to extend this program to more than two data sets, the program could become long and tedious. A macro approach can be used to accomplish the ID checking task with an arbitrary number of data sets. The next section demonstrates such a program.

Checking for an ID in Each of *n* Files

Data set Three is added to the mix to demonstrate how to approach this problem when there are more than two data sets. First, run Program 10.3 to create the new data set Three.

Program 10.3: Creating a Third Data Set for Testing Purposes

```
data Three;
   input Patno Gender $;
datalines;
1 M
2 F
3 M
5 F
6 M
12 M
13 M
;
```

Here is a listing of data set Three:

Figure 10.3: Listing of Data Set Three

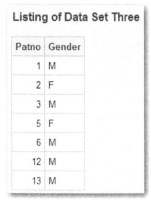

Listing of Data Set Three

Patno	Gender
1	M
2	F
3	M
5	F
6	M
12	M
13	M

Some of the observations in data set Three are in data set One, some in Two, and some not in either.

Before developing a macro, let's look at Program 10.4, which is a rather simple but slightly tedious program to accomplish the ID checks.

Program 10.4: Checking for an ID in Each of Three Data Sets (Long Way)

```
proc sort data=one(keep=Patno) out=Tmp1;
   by Patno;
run;

proc sort data=two(keep=Patno) out=Tmp2;
   by Patno;
run;

proc sort data=three(keep=Patno) out=Tmp3;
   by Patno;
run;

title "Listing of missing ID's and data set names";
data _null_;
   file print;
   merge Tmp1(in=In_Tmp1)
         Tmp2(in=In_Tmp2)
         Tmp3(in=In_Tmp3)  end=Last;
   by Patno;

   if not In_Tmp1 then do;
      put "ID " Patno "missing from data set One";
      n + 1;
   end;

   if not In_Tmp2 then do;
      put "ID " Patno "missing from data set Two";
      n + 1;
   end;

   if not In_Tmp3 then do;
      put "ID " Patno "missing from data set Three";
      n + 1;
   end;

   if Last and n eq 0 then
      put "All id's match in all files";
run;
```

Program 10.4 can be extended to accommodate any number of data sets. The output is shown next:

Figure 10.4: Output from Program 10.4

```
Listing of missing ID's and data set names

ID 2 missing from data set Two
ID 4 missing from data set One
ID 4 missing from data set Three
ID 6 missing from data set One
ID 6 missing from data set Two
ID 12 missing from data set Two
ID 13 missing from data set One
```

Notice that the PROC SORTS and IF statements follow a pattern that can be automated by writing a macro. The ID checking macro is developed in the next section.

A Macro for ID Checking

If you need to check that a given ID is present in each of *n* files and *n* is large, the DATA step approach above becomes too tedious. Using the same logic as Program 10.4, you can create a macro that will check for IDs across as many files as you wish.

Program 10.5: Presenting a Macro to Check for IDs Across Multiple Data Sets

```
*Program Name: Check_ID.sas
 Purpose: Macro which checks if an ID exists in each of n files
 Arguments: The name of the ID variable, followed by as many
            data sets names as desired, separated by BLANKS
 Example: %Check_ID(ID = Patno,
                    Dsn_list=One Two Three);

%macro Check_ID(ID=,        /* ID variable              */
               Dsn_list=   /* List of data set names,  */
                           /* separated by spaces      */);
   %do i = 1 %to 99;
     /* break up list into data set names */
     %let Dsn = %scan(&Dsn_list,&i,' ');
     %if &Dsn ne %then %do; /* If non null data set name      */
        %let n = &i;        /* When you leave the loop, n will */
                            /* be the number of data sets      */
        proc sort data=&Dsn(keep=&ID) out=Tmp&i;
           by &ID;
        run;
     %end;
   %end;

   title  "Report of data sets with missing ID's";
   data _null_;
      file print;
      merge
```

```
        %do i = 1 %to &n;
            Tmp&i(in=In_Tmp&i)
        %end;


        end=Last;
        by &ID;

        if Last and n eq 0 then do;
            put "All ID's Match in All Files";
            stop;
        end;

        %do i = 1 %to &n;
            %let Dsn = %scan(&Dsn_list,&i);
            if not In_Tmp&i then do;
                put "ID " &ID "missing from data set &dsn";
                n + 1;
            end;
        %end;


        run;
%mend Check_ID;
```

The two arguments used in this macro are the name of the ID variable and a list of data set names (separated by blanks). The first step is to break up the list of data set names into individual names. This is accomplished with the %SCAN function, which is similar to the non-macro SCAN function. It breaks strings into "words," where words are any string of characters separated by one or more delimiters. The three arguments to the %SCAN function are:

1. the string to be broken apart (parsed)
2. which "word" you want
3. an optional argument that lets you specify one or more delimiters.

In this program, a blank is used as the delimiter. This function will extract the individual data set names and stop when it runs out of data set names. For each of the data sets you obtain with the %SCAN function, you use PROC SORT to sort the data set by the ID variable and create a set of temporary data sets (Tmp1, Tmp2, etc.). When you run out of data set names, the %SCAN function returns a missing value and the program continues to the DATA _NULL_ step. The remainder of the program is similar to the previous program, where a MERGE statement is constructed using the multiple data set names followed by the IN= data set option.

To invoke this macro to check the three data sets (One, Two, and Three) with the ID variable Patno, you would write:

```
%check_ID(ID=Patno, Dsn_List=One Two Three)
```

The resulting output is the same as the output produced by Program 10.4.

Conclusions

This chapter described how to determine if a specific value, such as an ID, is present in any number of data sets. You might find that the macro presented at the end of this chapter will come in handy when you need to check IDs across multiple files. For just about any large project, you will most likely need to develop rules relating values from multiple data sets. A thorough knowledge of merging data sets and/or SQL will be necessary for these types of tasks.

Chapter 11: Using PROC COMPARE to Perform Data Verification

Introduction

Many critical data applications require that you enter the data twice and then compare the resulting files for discrepancies. This is usually referred to as *double entry* and *verification*. In the "old days," when I was first learning to use computers, most data entry was done using a keypunch (although my boys will tell you that, in my day, it was done with a hammer and chisel on stone tablets). The most common method of double entry and verification was done on a special keypunch machine called a *verifier*. The original deck of punched cards was placed in the input hopper, and a keypunch operator (preferably not the one who entered the data originally) re-keyed the information from the data entry form. If the information being typed matched the information already punched on the card, the card was accepted and a punch was placed, usually in column 81 of the card. If the information did not match, a check could be made to see whether the error was on the original card or in the re-keying of the information.

Today, there are several programs that accomplish the same goal by having all the data entered twice and then comparing the resulting data files. Some of these programs are quite sophisticated and also quite expensive. SAS software has a very flexible procedure called PROC COMPARE that can be used to compare the contents of two SAS data sets. You can refer to the *Base SAS 9.4 Procedures Guide, Sixth Edition* in hard copy or online documentation for more information. This chapter presents some simple examples using PROC COMPARE.

Conducting a Simple Comparison of Two Data Files

The simplest application of PROC COMPARE is presented first, determining if the contents of two SAS data sets are identical. Suppose that two people enter data from some coding forms, and the two data files are called File_1.txt and File_2.txt. A listing of the two files is shown next:

Figure 11.1: Double Entry Files File_1.txt and File_2.txt

```
File_1.txt

001M10211946130 80
002F12201950110 70
003M09141956140 90
004F10101960180100
007m10321940184110


File_2.txt

001M1021194613080
002F12201950110 70
003M09141956144 90
004F10101960180100
007M10231940184110
```

Here is the file format:

Variable	Description	Starting Column	Length	Type
Patno	Patient Number	1	3	Numeric
Gender	Gender	4	1	Character
DOB	Date of Birth	5	8	*mmddyyyy*
SBP	Systolic Blood Pressure	13	3	Numeric
DBP	Diastolic Blood Pressure	16	3	Numeric

The data, without mistakes, should have been:

Figure 11.2: Correct Data Representation

```
001M10211946130 80
002F12201950110 70
003M09141956140 90
004F10101960180100
007M10231940184110
```

A visual inspection of the two original files shows the following discrepancies:

For patient 001, there is a space missing before the 80 at the end of the line in File_2.txt.

For patient 003, SBP is 144 instead of 140 in File_2.txt.

For patient 007, the gender is entered in lowercase and the digits are interchanged in the day field of the date in File_1.txt.

Let's see how to use PROC COMPARE to detect these differences. You have some choices: One way to proceed is to create two SAS data sets, as shown in Program 11.1:

Program 11.1: Creating Data Sets One and Two from Two Raw Data Files

```
data One;
   infile "c:\Books\Clean3\File_1.txt" truncover;
   input @1  Patno  3.
         @4  Gender $1.
         @5  DOB    mmddyy8.
         @13 SBP    3.
         @16 DBP    3.;
   format DOB mmddyy10.;
run;

data Two;
   infile "c:\Books\Clean\File_2.txt" truncover;
   input @1  Patno  3.
         @4  Gender $1.
         @5  DOB    mmddyy8.
         @13 SBP    3.
         @16 DBP    3.;
   format DOB mmddyy10.;
run;
```

Then run PROC COMPARE, as shown in Program 11.2:

Program 11.2: Running PROC COMPARE

```
title "Using PROC COMPARE to Compare Two Data Sets";
proc compare base=One compare=Two;
   id Patno;
run;
```

The procedure options BASE= and COMPARE= identify the two data sets to be compared. In this example, data set One was arbitrarily chosen as the base data set. (The option DATA= can be used in place of BASE= because they are equivalent.)

The variable listed on the ID statement must be a unique identifier or a combination of variables that create a unique identifier.

Here is a partial listing of the output from PROC COMPARE:

Figure 11.3: Partial Output from Program 11.2

Using PROC COMPARE to Compare Two Data Sets

```
The COMPARE Procedure
Comparison of WORK.ONE with WORK.TWO
(Method=EXACT)

Data Set Summary

Dataset           Created          Modified    NVar    NObs

WORK.ONE  03DEC16:11:41:27  03DEC16:11:41:27     5       5
WORK.TWO  03DEC16:11:41:27  03DEC16:11:41:27     5       5


Variables Summary

Number of Variables in Common: 5.
Number of ID Variables: 1.
```

The first part of the output shows information about the two SAS data sets (data set names, date of creation, and date of modification), along with the number of variables the two data sets have in common and the number of observations in each data set.

Following this are several pages of summary information about differences between the two files. However, the more important part of the output is at the end, as shown:

```
Value Comparison Results for Variables


          ||  Base Value         Compare Value
  Patno   ||  Gender              Gender
────────  ||  ─                   ─
          ||
     7    ||  m                   M



          ||       Base     Compare
  Patno   ||       DOB          DOB        Diff.      % Diff
────────  ||  ─────────    ─────────    ─────────   ─────────
          ||
     7    ||       .       10/23/40         .           .



          ||       Base     Compare
  Patno   ||       SBP          SBP        Diff.      % Diff
────────  ||  ─────────    ─────────    ─────────   ─────────
          ||
     3    ||  140.0000     144.0000       4.0000      2.8571
```

This last part of the output shows all the differences between the two data sets. These differences are listed in the order that the variables appear in the SAS data sets. You will see in a moment how to instruct PROC COMPARE to present the differences in ID order.

Notice that the left-adjusted value of 80 for patient 001 in File_1.txt was not flagged as an error. Why? Because SAS correctly reads left-adjusted numeric values, and the comparison is between the two SAS data sets, not the raw files themselves. Also, the incorrect date of 10/32/1940 (patient number 005 in File_1) was shown as a missing value in the output. If you inspect the SAS Log, you will see that the incorrect date was flagged as an error. When invalid dates are encountered, SAS substitutes a missing value for that date. If you do not want this to happen, you can use a character informat instead of a date informat for data checking.

There are two useful options that you will want to use with PROC COMPARE. One is the BRIEF option, which reduces the output considerably by producing only a short comparison summary and suppressing the lengthy output shown in Figure 11.3. The other option is TRANSPOSE. This option places the error report in ID order rather than in variable order. Below is Program 11.2 rewritten with these two options added:

Program 11.3: Adding the BRIEF and TRANSPOSE Options to PROC COMPARE

```
title "Using PROC COMPARE to Compare Two Data Sets";
proc compare base=One compare=Two brief transpose;
   id Patno;
run;
```

The output, with these two options included, is shown in Figure 11.4:

Figure 11.4: Output from Program 11.3

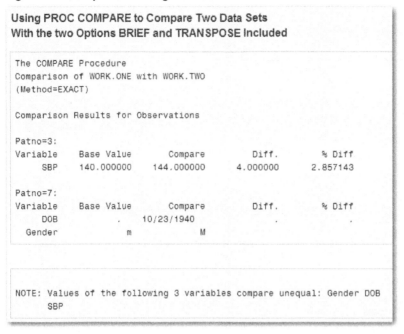

```
Using PROC COMPARE to Compare Two Data Sets
With the two Options BRIEF and TRANSPOSE Included

The COMPARE Procedure
Comparison of WORK.ONE with WORK.TWO
(Method=EXACT)

Comparison Results for Observations

Patno=3:
Variable    Base Value       Compare        Diff.       % Diff
    SBP     140.000000    144.000000     4.000000     2.857143

Patno=7:
Variable    Base Value       Compare        Diff.       % Diff
    DOB              .     10/23/1940            .            .
  Gender             m              M


NOTE: Values of the following 3 variables compare unequal: Gender DOB
      SBP
```

Because of the BRIEF option, this output only shows this single page and because of the TRANSPOSE option, the listing is in patient number (observation) order.

Simulating Double Entry Verification Using PROC COMPARE

If you want to see all of the differences in the two files, including the leading blank in the diastolic blood pressure (DBP) for patient 001 and the incorrect date (rather than a missing value), you can read in all the values as character variables, like this:

Program 11.4: Reading All the Variables as Character and Running PROC COMPARE

```
data One;
   infile "c:\Books\Clean3\File_1.txt" truncover;
   input @1  Patno  $char3.
         @4  Gender $char1.
         @5  DOB    $char8.
         @13 SBP    $char3.
         @16 DBP    $char3.;
run;

data Two;
   infile "c:\Books\Clean3\File_2.txt" truncover;
   input @1  Patno  $char3.
         @4  Gender $char1.
         @5  DOB    $char8.
         @13 SBP    $char3.
```

```
          @16 DBP        $char3.;
run;


title "Using PROC COMPARE to Compare Two Data Sets";
proc compare base=One compare=Two brief transpose;
    id Patno;
run;
```

Remember that the $CHAR*n*. informat does not right-justify character values compared to the $*n*. informat, which does. The output (shown in Figure 11.5) now includes all the differences between the two files, including the value for DBP in patient 001. You can also see both dates as they existed in the raw data files.

Figure 11.5: Output from Program 11.4

```
Using PROC COMPARE to Compare Two Data Dets


The COMPARE Procedure
Comparison of WORK.ONE with WORK.TWO
(Method=EXACT)


Comparison Results for Observations


Patno=001:
Variable     Base Value        Compare
    DBP             80              80


Patno=003:
Variable     Base Value        Compare
    SBP            140             144


Patno=007:
Variable     Base Value        Compare
  Gender            m               M
    DOB       10321940        10231940




NOTE: Values of the following 4 variables compare unequal: Gender DOB
      SBP DBP
```

Other Features of PROC COMPARE

PROC COMPARE can also be used with data sets that do not have the same number of observations or even the exact same set of variables. In the former case, you must include an ID variable and the comparison is performed on IDs that exist in both files. In the case where there are some variables in one data set that are not in the other, PROC COMPARE will automatically compare only those variables that are in both data sets. Finally, you can include a VAR statement with PROC COMPARE to list the variables from the two data sets that you want to include in the comparison.

Conclusions

PROC COMPARE is really quite an amazing procedure. You can use it to simulate double entry and verification as well as to compare two SAS data sets. I have been told that some companies use PROC COMPARE when they make a change to a SAS program and want to be sure that the SAS data set produced by the modified program is identical to the one produced by the original program. One other situation where PROC COMPARE is used is to guarantee that SAS data sets are equivalent when a company installs a newer SAS version.

Chapter 12: Correcting Errors

Introduction

Now that you have developed techniques for detecting data errors, it's time to think about how to correct those errors. The correcting process will vary, depending on your application. For example, you might have a large data warehouse where you can delete observations with errors (or possible errors) or replace possible data errors with missing values. For other applications, you might need to return to the original data forms, determine the correct values, and replace incorrect data values with corrected values.

There are a large number of errors in the 101 observations that make up the Patients data set. These errors were introduced so that various data cleaning methods could be demonstrated. It would be hoped that a small data set similar to the Patients data set would contain fewer errors.

Hard Coding Corrections

For small data sets with only a few corrections to be made, you might consider hard coding your corrections. As an example, take a look at Program 12.1:

Program 12.1: Hard Coding Corrections Using a DATA Step

```
*This program corrects errors in the Patients data set;

data Clean.Patients_01Jan2017;
   set Clean.Patients;

   ***Change lowercase values to uppercase;
   array Char_Vars[4] Patno Accoubt_No Gender Dx;
   do i = 1 to 4;
      Char_Vars[i] = upcase(Char_Vars[i]);
   end;
```

```
    if Patno = '003' then SBP = 110;
    else if Patno = '011' then Dx = '530.100';
    else if Patno ='023' then do;
       SBP = 146;
       DBP = 98;
    end;
    else if Patno = '034' then HR = 80;

***and so forth;
    drop i;
run;
```

This may seem somewhat inelegant (and it is), but if you need to make a few corrections, this method is reasonable. There are a couple of points you should notice about this program. First, you name the new data set to include a revision date, so that you can keep track of your data sets if you make more corrections in the future. Also, you will want to save this program with a name such as Update_01Jan2017.sas so that you can re-create the corrected data set from the original data should that need arise. **Never overwrite your original data set**. It is also strongly recommended that you insert a comment to document the source of your data changes. In addition to correcting several values, this program also converts each of the four character variables to uppercase.

Describing Named Input

You are most likely familiar with three ways of reading raw data in a SAS data set: list input (used for delimited data), column input (where you specify the starting and ending columns for each variable), and formatted input (where you use pointers and informats). There is one other method for reading raw data called *named input*. This input method is somewhat like the middle pedal on a piano—it's always been there, but nobody knows what it does or how to use it!

The reason that we are introducing this input method will become clear a little later in this chapter. The best way to explain named input is to show you an example. Take a look at Program 12.2:

Program 12.2: Describing Named Input

```
data Named;
   length Char $ 3;
   informat Date mmddyy10.;
   input x=
         y=
         Char=
         Date=;
datalines;
x=3 y=4 Char=abc Date=10/21/2010
y=7
Date=11/12/2016 Char=xyz x=9
;
```

To use named input, you include an equal sign after each variable name in the INPUT statement. This input method also requires you to include a variable name, an equal sign, and the value to be read in the raw data file.

In order to tell SAS that Char is a character variable with length 3 and that Date needs an *MMDDYY*10. informat, you use a LENGTH statement for the variable Char and an INFORMAT statement for the variable Date. (Note: You could have used the INFORMAT statement and assigned an informat of $3. for Char instead of using a LENGTH statement.)

You can enter your data in any order and if you leave out variables, they will automatically be given missing values. Please take a look at the listing below:

Figure 12.1: Listing of Data Set Named

Listing of Data Set Named

Char	Date	x	y
abc	18556	3	4
	.	.	7
xyz	20770	9	.

This input method is a bit cumbersome, but you will see that it can be very useful for correcting your data. Let's enter some (but not all) of the corrections to the Patients data set (described in Chapter 1) using named input.

Program 12.3: Using Named Input to Make Corrections

```
data Corrections_01Jan2017;
    length Patno $ 3
            Account_No Dx $ 7
            Gender $ 1;
    informat Visit mmddyy10.;
    format Visit date9.;
    input Patno=
            Account_No=
            Gender=
            Visit=
            HR=
            SBP=
            DBP=
            Dx=
            AE=;
datalines;
Patno=003 SBP=110
Patno=023 SBP=146 DBP=98
Patno=027 Gender=F
Patno=039 Account_No=NJ34567
Patno=041 Account_No=CT13243
Patno=045 HR=90
;
```

You use a LENGTH statement for the variables Patno, Account_No, Dx, and Gender. This sets the appropriate lengths for these variables and declares them to be character. An INFORMAT statement indicates that the visit date is to be read with the *MMDDYY*10. informat. Remember, you still need to format the variable Visit. Here the Date9 format is used. The listing is shown below:

Figure 12.2: Listing of Data Set Corrections_01Jan2017

Listing of Data Set Corrections_01Jan2017

Patno	Account_No	Dx	Gender	Visit	HR	SBP	DBP	AE
003				.	.	110	.	.
023				.	.	146	98	.
027			F	.	.	.	.	.
039	NJ34567			.	.	.	.	.
041	CT13243			.	.	.	.	.
045				.	90	.	.	.

In this example, you can see the advantage of using named input. Rather than having to place your data values in columns or enter periods for missing values, you can simply enter values for the variables you want to change. The question is, what do you do with this data set? Let's take a moment to review the UPDATE statement.

Reviewing the UPDATE Statement

If you merge two data sets that share variable names, the value from the right-most data set will replace the value from the data set to its left, even if the value in the right-most data set is a missing value. However, if you use an UPDATE statement instead of a MERGE statement, a missing value for a variable in the right-most data set will not replace the value in the data set to its left. The right-most data set is sometimes called the *transaction data set*. It is used to update values in the master data set. To be sure this is clear, we will demonstrate how the UPDATE statement works with two small data sets, Inventory and Transaction. Take a look:

Program 12.4: Demonstrating How the UPDATE Statement Works

```
data Inventory;
   length PartNo $ 3;
   input PartNo $ Quantity Price;
datalines;
133 200 10.99
198 105 30.00
933 45 59.95
;

data Transaction;
   length PartNo $ 3;
   input Partno=
         Quantity=
         Price=;
```

```
datalines;
PartNo=133 Quantity=195
PartNo=933 Quantity=40 Price=69.95
;
proc sort data=Inventory;
    by Partno;
run;

proc sort data=Transaction;
    by PartNo;
run;

data Inventory_22Feb2017;
    update Inventory Transaction;
    by Partno;
run;
```

To help visualize how the UPDATE statement is working, here is a listing of the Inventory and Transaction data sets (after the sort):

Figure 12.3: Listing of Data Sets Inventory and Transaction

Listing of Data Set Inventory

PartNo	Quantity	Price
133	200	10.99
198	105	30.00
933	45	59.95

Listing of Data Set Transaction

PartNo	Quantity	Price
133	195	.
933	40	69.95

The Transaction data set contains new values for two of the three part numbers in the Inventory data set (OK, it's a pretty small company). You have an updated quantity for part number 133 and a new quantity and price for part number 933. Below is a listing of the updated data set:

Figure 12.4: Updated Inventory Data Set Inventory_22Feb2017

Listing of Data Set Inventory_22Feb2017

PartNo	Quantity	Price
133	195	10.99
198	105	30.00
933	40	69.95

Whatever method you decide to use when correcting your data, make sure that you are methodical. You should be able to re-create the final, corrected data set from the raw data at any time.

> Note: When using the UPDATE statement, the master data set must have unique values of the BY value for all observations. If there are duplicates, only the first one is updated by the transaction data set.

Using the UPDATE Statement to Correct Errors in the Patients Data Set

Before we leave this chapter, let's correct the errors in the Patients data set by creating a transaction file containing corrected data. Because of the large number of errors, you might choose to use named input to make a transaction file and then use the UPDATE statement to make the corrections. The transaction file, listed next, contains corrected data for all of the errors detected by the methods described in the previous chapters. The new values, just like all the original values, are made up. Here is the program:

Program 12.5: Creating the Transaction Data Set to Clean the Patients Data Set

```
data Corrections_01Jan2017;
   length Patno $ 3
          Account_No Dx $ 7
          Gender $ 1;
   informat Visit mmddyy10.;
   ***Note: The MMDDYY10. format is used here to be compatible
      with the original file;
   format Visit mmddyy10.;
   input Patno=
         Account_No=
         Gender=
         Visit=
         HR=
         SBP=
         DBP=
         Dx=
         AE=;
datalines;
Patno=003 SBP=110
Patno=009 Visit=03/15/2015
Patno=011 Dx=530.100
Patno=016 Visit=10/21/2016
Patno=023 SBP=146 DBP=98
Patno=027 Gender=F
Patno=039 Account_No=NJ34567
Patno=041 Account_No=CT13243
Patno=045 HR=90
Patno=050 HR=32
Patno=055 Gender=M
Patno=058 Gender=M
Patno=088 Gender=F
Patno=094 Dx=023.000
Patno=095 Gender=F
Patno=099 DBP=60
;

proc sort data= Corrections_01Jan2017;
   by Patno;
run;
```

Here is a listing of the data set containing the corrections:

Figure 12.5: Listing of Data Set Corrections_01Jan2017

Listing of Data Set Corrections_01Jan2017

Patno	Account_No	Dx	Gender	Visit	HR	SBP	DBP	AE
003				.	.	110	.	.
009				03/15/2015	.	.	.	.
011		530.100		.	.	.	.	.
016				10/21/2016	.	.	.	.
023				.	.	146	98	.
027			F	.	.	.	.	.
039	NJ34567			.	.	.	.	.
041	CT13243			.	.	.	.	.
045				.	90	.	.	.
050				.	32	.	.	.
055			M	.	.	.	.	.
058			M	.	.	.	.	.
088			F	.	.	.	.	.
094		023.000		.	.	.	.	.
095			F	.	.	.	.	.
099				.	.	.	60	.

This data set contains values for Patno and non-missing values for all variables that need to be corrected.

Before you can use the UPDATE statement to apply the corrections to this data set, you first need to do several things: One, you need to remove the duplicate record for 005. Next, you need to correct the two duplicate patient numbers (that represent data for different subjects). Finally, the patient with the number XX5 and the patient with a missing patient number need to be assigned unique patient numbers. Program 12.6 accomplishes these goals:

Program 12.6: Taking Care of Duplicate Records, Duplicate Patient Numbers, and Incorrect Patient Numbers

```
*First remove the duplicate observation in the Patients data set
 before performing the update;

proc sort data=Clean.Patients out=Patients_No_Duprecs noduprecs;
   by Patno;
run;

*Next fix the two patients with the same value of Patno but with
 different data. Also correct the incorrect patient number 'XX5' and the
missing patient number;
```

```
data Fix_Incorrect_Patno;
   set Patients_No_Duprecs;

   ***Correct duplicate patient numbers;
   if Patno='007' and Account_no='NJ90043' then Patno='102';
   else if Patno='050' and Account_No='NJ87682' then Patno='103';

   ***Correct incorrect and missing patient numbers;
   if Patno='XX5' then Patno='101';
   ***There was only one missing patient number;
   if missing(Patno) then Patno='104';
run;

proc sort data=Fix_Incorrect_Patno;
   by Patno;
run;
```

To remove the duplicate observation for patient 005, you use PROC SORT with the NODUPRECS option. The other corrections are hard coded. You finish this step by sorting the corrected data set by Patno.

You are now ready to use the UPDATE statement with the transaction data set to correct all the remaining errors in the data, as follows:

Program 12.7: Correcting Data with a Transaction Data Set and Uppercasing Character Variables

```
*Using the transaction data set to correct errors in the original data set
Fix_Incorrect_Patno;

data Clean.Patients_02Jan2017;
   update Fix_Incorrect_Patno Corrections_01Jan2017;
   by Patno;

   *Upcase all character variables;
   array Char_Vars[4] Patno Accoubt_No Gender Dx;
   do i = 1 to 4;
      Char_Vars[i] = upcase(Char_Vars[i]);
   end;
   drop i;
run;

title "Listing of Data Set Clean.Patients_02Jan2017";
proc print data=Clean.Patients;
   id Patno;
run;
```

The Corrections_01Jan2017 data set was previously sorted by Patno. The UPDATE statement is used to correct the various errors in the original data set. The last item to address is to uppercase the character variables. The final data set, Patients_02Jan2017, contains the final corrected data.

A partial listing of this corrected data set is shown below:

Figure 12.6: Partial Listing of Data Set Patients_02Jan2017

Listing of Data Set Clean.Patients_02Jan2017

Patno	Account_No	Gender	Visit	HR	SBP	DBP	Dx	AE
001	CT14882	M	06/12/2012	69	124	86	713.410	0
002	MD78461	M	06/04/2010	76	130	80	047.570	1
003	DE51381	f	06/22/2013	70	56	70	108.510	0
004	CT37146	M	05/18/2013	76	112	84	669.860	0
005	DE00080	F	04/08/2012	91	106	84	078.160	0
005	DE00080	F	04/08/2012	91	106	84	078.160	0
006	DE37709	M	07/27/2014	71	104	88	967.570	0
007	VT56383	F	01/13/2014	63	128	80	640.260	1
008	PA67069	F	09/28/2013	79	124	72	020.120	0
009	MD68313	F	03/15/1956	82	132	64	894.400	0
007	NJ90043	M	08/06/2010	83	130	102	564.870	0
011	NY60612	F	07/03/2010	68	110	78	530.abc	0
012	CT77620	M	08/21/2014	70	124	78	904.490	0
						92	985	

In this small section of the listing, you can see that some of the corrected values listed in the Corrections_01Jan2017 data set have replaced the incorrect values (other corrections are further down in the data set).

Once you have a clean data set, you probably want to keep it that way. As you will see in the next chapter, you can add integrity constraints to a data set. This can help prevent data errors from being added to the clean data set as new data is collected.

Conclusions

The two basic approaches to correcting data errors are: 1) if there are only a few errors, hard code the corrections; and 2) for larger data sets with more errors, create a transaction data set of the corrections (perhaps using named input) and use the UPDATE statement to replace the data errors with the correct values.

Chapter 13: Creating Integrity Constraints and Audit Trails

Introduction

Starting with Version 7 of SAS, a feature called integrity constraints was implemented (features of this system were added or improved in later versions). *Integrity constraints* are rules that are stored within a SAS data set that can restrict data values accepted into the data set when new data is added with PROC APPEND; the MODIFY statement in the DATA step; and SQL insert, delete, or update. These constraints are preserved when the data set is copied using PROC COPY, CPORT, or CIMPORT or when it is sorted with PROC SORT. The UPLOAD and DOWNLOAD procedures also preserve constraints.

There are two types of integrity constraints. One type, called *general integrity constraints*, allows you to restrict data values that are added to a single SAS data set. You can specify specific values for a variable or a range of values or you can specify a requirement that values of a variable be unique and/or non-missing. You can also create integrity constraints that reflect relationships among variables in a data set.

The other type, called *referential integrity constraints*, allows you to link data values among SAS data sets. For example, you might have a list of valid patient numbers in a demographic file. You could restrict any patient numbers added to another file to those that exist in the demographic file. You could also prevent the deletion of patient numbers in a demographic file unless all occurrences of those patient numbers were first removed from the related files. Finally, you could link the demographic file and one or more related files so that any change to a patient number in the demographic file would automatically change all the associated patient number values in the related files.

Integrity constraints can keep a data set "pure" by rejecting any observations that violate one or more of the constraints. You can have SAS create an audit trail data set, providing you with information about observations that failed to meet one or more of the constraints. Using referential constraints, you can ensure the consistency of variable values among multiple files.

Integrity constraints can be created with PROC SQL statements, PROC DATASETS, or SCL (SAS Component Language) programs. In this chapter, PROC DATASETS is used to create or delete integrity constraints. If you are already familiar with using SQL to manage integrity constraints, you can choose to use PROC SQL to create, delete, or manage your integrity constraints. Any SQL reference can give you details on the SQL approach. There is also some limited information on using PROC SQL to create and manage integrity constraints in SAS online documentation.

Demonstrating General Integrity Constraints

There are four types of general integrity constraints:

Check
A user-defined constraint on a single variable or a relationship among several variables. You can specify a list of values, a range of values, or a relationship among two or more variables.

Not Null
Missing values are not allowed for this variable.

Unique
Values for this variable must be unique.

Primary Key
Values for this variable (or variables) must be both unique and non-missing. This constraint is particularly useful for ID variables.

To demonstrate how you could use integrity constraints to prevent data errors from being added to an existing SAS data set, let's first create a small data set called Health by running the following program:

Program 13.1: Creating Data Set Health to Demonstrate Integrity Constraints

```
data Health;
   informat Patno $3. Gender $1.;
   input Patno Gender HR SBP DBP;
datalines;
001 M  88 140  80
002 F  84 120  78
003 M  58 112  .
004 F  66 200 120
007 M  88 148 102
015 F  82 148  88
;
```

A listing of this data set is shown next:

Figure 13.1: Listing of Data Set Health

Patno	Gender	HR	SBP	DBP
001	M	88	140	80
002	F	84	120	78
003	M	58	112	.
004	F	66	200	120
007	M	88	148	102
015	F	82	148	88

The next program adds general integrity constraints to the Health data set. The constraints are:

- Gender must be 'F' or 'M'.
- HR (heart rate) must be between 40 and 100. Missing values are not allowed.
- SBP (systolic blood pressure) must be between 50 and 240. Missing values are allowed.
- DBP (diastolic blood pressure) must be between 35 and 130. Missing values are allowed.
- Patno (patient number) must be unique and non-missing.

Here are the PROC DATASETS statements:

Program 13.2: Creating Integrity Constraints Using PROC DATASETS

```
proc datasets library=Work nolist;
   modify Health;
   ic create Gender_Chk = check
      (where=(Gender in('F','M')));

   ic create HR_Chk = check
      (where=(HR between 40 and 100));

   ic create SBP_Chk = check
      (where=(SBP between 50 and 240 or SBP is missing));

   ic create DBP_Chk = check
      (where=(DBP between 35 and 130 or DBP is missing));

   ic create ID_Chk = primary key(Patno);
run;
quit;
```

IC CREATE (stands for Integrity Constraint create) is followed by a name for the constraint. The name can be any valid SAS name; however, it makes sense to choose a name that is related to the purpose of the constraint. You can even give the constraint the same name as the variable you are referencing.

Note that if any of the observations in an existing data set violate any of the integrity constraints you want to create, that integrity constraint will not be created (in the example presented here, the existing data sets does not contain any IC violations). You have some choices: First, you can be sure that there are no integrity constraint violations in your data set before you run PROC DATASETS, or you can create a data set with all the variables but no observations (as shown in Program 13.11), make the constraints, and then use PROC APPEND to start adding data.

Running PROC CONTENTS will display the usual data set information as well as the integrity constraints. The resulting output (edited) from running PROC CONTENTS is shown next:

Figure 13.2: Partial Listing of PROC CONTENTS

#	Variable	Type	Len
5	DBP	Num	8
2	Gender	Char	1
3	HR	Num	8
1	Patno	Char	3
4	SBP	Num	8

Alphabetic List of Variables and Attributes

This section is the normal alphabetic list of variables.

Partial Listing of PROC CONTENTS (continued)

Alphabetic List of Integrity Constraints

#	Integrity Constraint	Type	Variables	Where Clause
1	DBP_Chk	Check		(DBP>=35 and DBP<=130) or (DBP is null)
2	Gender_Chk	Check		Gender in ('F', 'M')
3	HR_Chk	Check		(HR>=40 and HR<=100)
4	ID_Chk	Primary Key	Patno	
5	SBP_Chk	Check		(SBP>=50 and SBP<=240) or (SBP is null)

Alphabetic List of Indexes and Attributes

#	Index	Unique Option	Owned by IC	# of Unique Values
1	Patno	YES	YES	6

Notice that each of the WHERE clauses that created the integrity constraints are listed in the output from PROC CONTENTS.

What happens when you try to append data that violates one or more of the integrity constraints? The short DATA step shown next creates a data set (New) containing four observations:

Program 13.3: Creating Data Set New Containing Valid and Invalid Data

```
data New;
    input Patno : $3. Gender : $1. HR SBP DBP;
datalines;
456 M 66 98 72
567 F 150 130 80
003 M 70 134 86
123 F 66 10 80
013 X . 120 90
;
```

Note: Values in bold represent integrity constraint violations. Here is a listing of data set New:

Figure 13.3: Listing of Data Set New

Listing of Data Set New

Patno	Gender	HR	SBP	DBP
456	M	66	98	72
567	F	150	130	80
003	M	70	134	86
123	F	66	10	80
013	X	.	120	90

Data for patient 456 is valid and consistent with all the constraints; patient 567 has a heart rate (HR) outside the valid range; patient 003 is a duplicate patient number; patient 123 has an SBP outside the valid range; and patient 013 has an invalid Gender and a missing value for HR, both of which violate the constraints.

Describing PROC APPEND

Before we go ahead and attempt to add the data from data set New to the Health data set, it might be a good idea to take a moment to review PROC APPEND. (Please skip to the next section if you already know this stuff.)

One way to concatenate two data sets is with s SET statement. Suppose that you have two data sets Big and Small, and you want to add the observations in Small to those in Big. You could do it this way:

```
data Combined;
    set Big Small;
run;
```

This is fine, except you need to read and write all the observations in Big before adding the observations in Small. As long as the data descriptors in the two data sets are the same (same variables and same lengths for the character variables), you can use PROC APPEND instead, like this:

```
proc append base=Big Data=Small;
run;
```

You specify the original data set with the BASE= option and the data set you want to append with the DATA= option. PROC APPEND goes to the end of data set Big (without reading any observations) and then adds the observations from data set Small. As a result, data set Big now contains the concatenated data from the original data set Big and the observations from data set Small. This is a much more efficient way of concatenating data sets; however, this method is analogous to burning your bridges behind you because you can't undo the changes. If you have any uncertainty in this process, you might want to make a copy of Big before you run PROC APPEND (at least for the first time). The reason that it is important for the variables in data set Small to match those in data set Big is that the data descriptor is taken from data set Big. If the attributes (such as the length of character variables) are not the same in the two data sets being concatenated, PROC APPEND will not run (unless you use an option called FORCE—and that is dangerous). Now, back to integrity constraints.

Demonstrating How Integrity Constraints Block the Addition of Data Errors

Let's see what happens when you attempt to append the observations in data set New to the original Health data set:

Program 13.4: Attempting to Append Data Set New to the Health Data Set

```
proc append base=Health data=New;
run;
```

Program 13.4 was run and a listing of the SAS Log is displayed in the figure below:

Figure 13.4: SAS Log from Running Program 13.4

```
163
164   proc append base=Health data=New;
165   run;

NOTE: Appending WORK.NEW to WORK.HEALTH.
WARNING: Add/Update failed for data set WORK.HEALTH because data value(s)
do not comply with integrity constraint ID_Chk,
1 observations rejected.
WARNING: Add/Update failed for data set WORK.HEALTH because data value(s)
do not comply with integrity constraint Gender_Chk,
1 observations rejected.
WARNING: Add/Update failed for data set WORK.HEALTH because data value(s)
do not comply with integrity constraint SBP_Chk, 1
observations rejected.
WARNING: Add/Update failed for data set WORK.HEALTH because data value(s)
do not comply with integrity constraint HR_Chk,
1 observations rejected.
NOTE: There were 5 observations read from the data set WORK.NEW.
```

```
NOTE: 1 observations added.
NOTE: The data set WORK.HEALTH has 7 observations and 5 variables.
```

The log shows that only one observation from data set New was appended to the Health data set. In addition, you see that one observation failed the HR_Chk constraint, one observation failed the SBP_Chk constraint, one observation failed the Gender_Chk constraint, and one observation failed the ID_Chk constraint. Even though patient 013 failed two constraints (invalid Gender and a missing value for HR), only one of the failed constraints was reported. SAS only reports the first integrity constraint (in the order they were created) for each observation.

Adding Your Own Messages to Violations of an Integrity Constraint

Messages such as the ones in the SAS Log above are not all that useful unless you have a listing of each constraint and, even then, it is tedious to look up each constraint violation. The solution is to write your own message for each constraint violation. Here is Program 13.2, rewritten to include user messages:

Program 13.5: User Messages Included in the Integrity Constraints

```
proc datasets library=Work nolist;
   modify Health;
   ic create Gender_Chk = check
      (where=(Gender in('F','M')))
      message="Gender must be F or M"
      msgtype=user;

   ic create HR_Chk = check
      (where=(HR between 40 and 100))
      message="HR must be between 40 and 100"
      msgtype=user;

   ic create SBP_Chk = check
      (where=(SBP between 50 and 240 or SBP is missing))
      message="SBP must be between 50 and 240 or missing"
      msgtype=user;

   ic create DBP_Chk = check
      (where=(DBP between 35 and 130 or DBP is missing))
      message="DBP must be between 35 and 130 or missing"
      msgtype=user;

   ic create ID_Chk = primary key (Patno)
      message="Patno must be unique and non-missing"
      msgtype=user;
run;
quit;
```

You write your own message following the keyword MESSAGE= and include a message type (MSGTYPE=) of USER. Adding a message and a message type equal to user results in your messages being printed instead of the default messages.

With the user messages included, let's see the result of appending data set New to data set Health (it was returned to its original state). The log now looks like this:

Figure 13.5: SAS Log Containing User Messages

```
45    proc append base=Health data=New;
46    run;

NOTE: Appending WORK.NEW to WORK.HEALTH.
WARNING: Patno must be unique and non-missing , 1 observations rejected.
WARNING: Gender must be F or M , 1 observations rejected.
WARNING: SBP must be between 50 and 240 or missing , 1 observations
rejected.
WARNING: HR must be between 40 and 100 , 1 observations rejected.
NOTE: There were 5 observations read from the data set WORK.NEW.
NOTE: 1 observations added.
NOTE: The data set WORK.HEALTH has 7 observations and 5 variables.
```

It is clear that it is worth your time to include user messages in all your integrity constraints.

Deleting an Integrity Constraint Using PROC DATASETS

You can use PROC DATASETS to delete one or more integrity constraints. For example, if you want to remove the constraint on Gender from the Health data set, you would proceed like this:

Program 13.6: Deleting an Integrity Constraint Using PROC DATASETS

```
proc datasets library=Work nolist;
   modify Health;
   ic delete Gender_Chk;
run;
quit;
```

Creating an Audit Trail Data Set

You might have a large data warehouse where you can simply throw out observations that fail one or more integrity constraints. If, however, you want to keep track of observations that failed one or more integrity constraints, you need to create an audit trail data set. An audit trail data set has the same name as your SAS data set with a type of audit. To see the contents of an audit trail data set, say in a PROC PRINT or a PROC REPORT, you need to use the TYPE=AUDIT data set option. You will see this demonstrated in a moment.

Now it's time to create the audit trail data set. Program 13.7 (below) uses PROC DATASETS to do this:

Program 13.7: Creating an Audit Trail Data Set

```
proc datasets library=Work nolist;
   audit Health;
   initiate;
run;
quit;
```

Program 13.7 creates an audit trail data set that will log all attempts to add data to the Health data set. Besides INITIATE, once you have created the audit trail data set, you can SUSPEND (suspend event logging but do not delete the audit trail data set), RESUME (resume event logging), and TERMINATE (terminate event logging and delete the audit trail data set).

Let's rerun Program 13.4 and inspect the contents of the audit trail data set using PROC PRINT. (As before, data set Health was returned to its original condition with six observations and no errors.)

Program 13.8: Using PROC PRINT to List the Contents of the Audit Trail Data Set

```
title "Listing of the Audit Trail Data Set";
proc print data=Health(type=audit) noobs;
run;
```

Here is the listing:

Figure 13.6: Listing the Audit Trail Data Set Using PROC PRINT

Listing of the Audit Trail Data Set

Patno	Gender	HR	SBP	DBP	_ATDATETIME_	_ATOBSNO_	_ATRETURNCODE_
456	M	66	98	72	10DEC2016:10:40:34	7	.
567	F	150	130	80	10DEC2016:10:40:34	8	.
003	M	70	134	86	10DEC2016:10:40:34	9	.
123	F	66	10	80	10DEC2016:10:40:34	10	.
013	X	.	120	90	10DEC2016:10:40:34	11	.
567	F	150	130	80	10DEC2016:10:40:34	8	-2147348311
123	F	66	10	80	10DEC2016:10:40:34	10	-2147348311
013	X	.	120	90	10DEC2016:10:40:34	11	-2147348311
003	M	70	134	86	10DEC2016:10:40:34	9	-2147348311

ATUSERID	_ATOPCODE_	_ATMESSAGE_
Ron's Dell	DA	
Ron's Dell	DA	
Ron's Dell	DA	
Ron's Dell	DA	
Ron's Dell	DA	
Ron's Dell	EA	ERROR: HR must be between 40 and 100
Ron's Dell	EA	ERROR: SBP must be between 50 and 240 or missing
Ron's Dell	EA	ERROR: Gender must be F or M
Ron's Dell	EA	ERROR: Patno must be unique and non-missing

Remember that the audit trail data set has the same name as your SAS data set, so you must use the TYPE= data set option. Inspection of the listing shows the original five observations in data set New, followed by the four observations in data set New that included one or more integrity constraint errors. The _ATOPCODE_ variable has a value of 'EA' (Error Add) for these observations, the _ATRETURNCODE_ variable is non-missing, and the _ATMESSAGE_ variable contains the user message you created with the integrity constraints.

Other automatic variables that you might find useful in the audit trail data set are _ATDATETIME_ (which reports the date and time a change was made) and _ATUSERID_ (which reports who made the changes).

To see only those observations that contained one or more integrity constraint violations, you can select observations where the value of _ATOPCODE_ is equal to 'EA' (error adding), 'ED' (error deleting), or 'EU' (error updating). Below is a list of all the _ATOPCODE_ values:

Table 13.1: Listing of _ATOPCODE_ Values

Code	Modification
AL	Auditing is resumed
AS	Auditing is suspended
DA	Added data record image
DD	Deleted data record image
DR	Before-update record image
DW	After-update record image
EA	Observation add failed
ED	Observation delete failed
EU	Observation update failed

Armed with this information, you can use PROC REPORT (or PROC PRINT if you prefer) to create an error report, as follows:

Program 13.9: Reporting the Integrity Constraint Violations Using the Audit Trail Data Set

```
title "Integrity Constraint Violations";
proc report data=Health(type=audit);
   where _ATOPCODE_ in ('EA' 'ED' 'EU');
   columns Patno Gender HR SBP DBP _ATMESSAGE_;
   define Patno / order "Patient Number" width=7;
   define Gender / display width=6;
   define HR / display "Heart Rate" width=5;
   define SBP / display width=3;
   define DBP / display width=3;
   define _atmessage_ / display "_IC Violation_"
                        width=30 flow;
run;
```

Here is the final report:

Figure 13.7: Output from Program 13.10

Integrity Constraint Violations

Patient Number	Gender	Heart Rate	SBP	DBP	IC Violation
003	M	70	134	86	ERROR: Patno must be unique and non-missing
013	X	.	120	90	ERROR: Gender must be F or M
123	F	66	10	80	ERROR: SBP must be between 50 and 240 or missing
567	F	150	130	80	ERROR: HR must be between 40 and 100

This is looking much better. The only (slight) problem is that if there is more than one integrity constraint violation in an observation, you will only see the first one (based on the order that they were created). For example, patient 013 had an invalid value for Gender ('X') and HR is missing. The _ATMESSAGE_ variable listed the Gender violation. Therefore, you could correct the violation that is listed in the audit trail data set and still have the observation fail an additional integrity constraint.

If you create a data set consisting only of the observations containing integrity constraint errors (as in the report above), you can correct the reported errors and attempt to run PROC APPEND again. Then, if there are still uncorrected errors, you can continue to run PROC APPEND until all the observations that you want to add have been corrected.

As an example, we will correct each of the reported errors in the listing above as follows:

Program 13.10: Correcting Errors Based on the Observations in the Audit Trail Data Set

```
data Correct_Audit;
   set Health(type=Audit
              where=(_ATOPCODE_ in ('EA' 'ED' 'EU')));
   if Patno = '003' then Patno = '103';
   else if Patno = '013' then do;
      Gender = 'F';
      HR = 88;
   end;
   else if Patno = '123' then SBP = 100;
   else if Patno = '567' then HR = 87;
   drop _AT: ;
run;

proc append base=Health data=Correct_Audit;
run;
```

In Program 13.10, all of the errors have been corrected and the corrected file was appended to the original Health data set. The notation _AT: in the DROP statement is an instruction to drop all variables beginning with _AT. Looking at the log (below), you notice that no integrity constraints were violated and that all four observations were added to the Health data set.

Figure 13.8: Listing of the SAS Log When Program 13.11 Is Run

```
173  proc append base=Health data=Correct_Audit;
174  run;

NOTE: Appending WORK.CORRECT_AUDIT to WORK.HEALTH.
NOTE: There were 4 observations read from the data set WORK.CORRECT_AUDIT.
NOTE: 4 observations added.
NOTE: The data set WORK.HEALTH has 11 observations and 5 variables.
```

Demonstrating an Integrity Constraint Involving More Than One Variable

Check integrity constraints can involve more than a single variable. Suppose that you have a survey where you ask how much time a person spends doing various activities, as a percentage. Each individual value cannot exceed 100%, and the sum of the values also cannot exceed 100%. The program below first creates a data set with all the required variables but with 0 observations (my thanks to Mike Zdeb, who suggested this method).

Program 13.11: Creating a Data Set with Variables and No Observations

```
data Survey;
   length ID $ 3;
   retain ID ' ' TimeTV TimeSleep TimeWork TimeOther .;
   stop;
run;
```

To convince you that this program really works as advertised, here is a listing of the SAS Log after the program was submitted:

Figure 13.9: Listing of the SAS Log After Program 13.11 Is Run

```
175  data Survey;
176     length ID $ 3;
177     retain ID ' ' TimeTV TimeSleep TimeWork TimeOther .;
178     stop;
179  run;
```

NOTE: The data set WORK.SURVEY has 0 observations and 5 variables.

Next, PROC DATASETS is used to create integrity constraints for each of the percentage variables as well as a constraint on the sum of these variables. The audit trail data set is created in this same procedure.

Program 13.12: Demonstrating an Integrity Constraint Involving More Than One Variable

```
proc datasets library=Work nolist;
   modify Survey;
   ic create ID_check = primary key(ID)
      message = "ID must be unique and non-missing"
      msgtype = user;

   ic create TimeTV_max = check(where=(TimeTV le 100))
      message = "TimeTV must not be over 100"
      msgtype = user;

   ic create TimeSleep_max = check(where=(TimeSleep le 100))
      message = "TimeSleep must not be over 100"
      msgtype = user;

   ic create TimeWork_max = check(where=(TimeWork le 100))
      message = "TimeWork must not be over 100"
      msgtype = user;

   ic create TimeOther_max = check(where=(TimeOther le 100))
      message = "TimeOther must not be over 100"
      msgtype = user;
```

```
   ic create Time_total =
      check(where=(sum(TimeTV,TimeSleep,TimeWork,TimeOther) le 100))
      message = "Total percentage cannot exceed 100%"
      msgtype = user;

run;

   audit Survey;
   initiate;
run;
quit;
```

Data set Survey has five variables and 0 observations. The data set (with or without observations) must exist before you can add integrity constraints.

The last integrity constraint (Time_total) prevents any observations where the sum of the percentages exceeds 100% from being added to the data set. When you create your data set of survey values and attempt to append it to data set Survey (which originally contained no observations), here is what happens:

Program 13.13: Adding the Survey Data

```
data Add;
   length ID $ 3;
   input ID $ TimeTV TimeSleep TimeWork TimeOther;
datalines;
001 10 40 40 10
002 20 50 40 5
003 10 . . .
004 0 40 60 0
005 120 10 10 10
;

proc append base=Survey data=Add;
run;

title "Integrity Constraint Violations";
proc report data=survey(type=audit);
   where _ATOPCODE_ in ('EA' 'ED' 'EU');
   columns ID TimeTV TimeSleep TimeWork TimeOther _ATMESSAGE_;
   define ID / order "ID Number" width=7;
   define TimeTV / display "Time spent watching TV" width=8;
   define TimeSleep / display "Time spent sleeping" width=8;
   define TimeWork / display "Time spent working" width=8;
   define TimeOther / display "Time spent in other activities"
                     width=10;
   define _atmessage_ / display "_Error Report_"
                     width=30 flow;
run;
```

The resulting report is shown next:

Figure 13.10: Audit Trail Error Report

Integrity Constraint Violations

ID Number	Time spent watching TV	Time spent sleeping	Time spent working	Time spent in other activities	Error Report
002	20	50	40	5	ERROR: Total percentage cannot exceed 100%
005	120	10	10	10	ERROR: TimeTV must not be over 100

None of the individual percentages for ID 002 exceeds 100% but, since the total of these percentages does exceed 100%, the Time_total integrity constraint is violated and that observation is not added to the survey. Notice that the message for ID 005 indicates that the value for TimeTV exceeds 100%, even though this observation also violates the total percentage constraint. Remember that you only see the message corresponding to the first integrity constraint violation for each observation.

Demonstrating a Referential Constraint

In the beginning of this chapter, two classes of constraints were described: General and Referential. *General constraints* relate to a single data set while *referential constraints* restrict certain operations on one or more variables among two or more data sets. You choose one or more variables (listed as a primary key) from one data set (called the *parent data set*) and create relationships with the same variable or variables (called foreign keys) in one or more data sets (called *child data sets*). The purpose is to restrict modifications to either the primary or foreign keys.

If you update or delete an observation in the parent data set, you need to specify what action you wish to take. Possible actions are:

RESTRICT

Prevents the values in the primary key variables from being updated or deleted in the parent data set if these same key values exist as a foreign key in a child data set. This is the default action if you do not specify an action.

SET NULL

Allows you to update or delete primary key values when there are matching values in a foreign key data set, but the values of the foreign keys are set to missing.

CASCADE

Allows values of the primary key variables to be changed, with corresponding values of the foreign keys changed to the same values.

It is important that the variables that you define as primary and foreign key variables are the same type and length.

As an example, we will first create two data sets, Master_List and Salary. Both data sets contain first name (FirstName) and last name (LastName) variables. In this scenario, be sure that you cannot change or delete a name in the Master_List data set if that name appears in the Salary data set. The program that follows creates the two data sets and sets up a referential constraint and two general constraints:

Program 13.14: Creating Two Data Sets and a Referential Constraint

```
data Master_List;
   informat FirstName LastName $12. DOB mmddyy10. Gender $1.;
   input FirstName LastName DOB Gender;
   format DOB mmddyy10.;
datalines;
Julie Chen 7/7/1988 F
Nicholas Schneider 4/15/1966 M
Joanne DiMonte 6/15/1983 F
Roger Clement 9/11/1988 M
;

data Salary;
   informat FirstName LastName $12. Salary comma10.;
   input FirstName LastName Salary;
datalines;
Julie Chen $54,123
Nicholas Schneider $56,877
Joanne DiMonte $67,800
Roger Clement $42,000
;

title "Listing of Master List";
proc print data=Master_List;
run;
title "Listing of Salary";
proc print data=Salary;
run;

proc datasets library=Work nolist;
   modify Master_List;
   ic create Prime_Key = primary key (FirstName LastName);
   ic create Gender_Chk = check(where=(Gender in ('F','M')));
run;

   modify Salary;
   ic create Foreign_Key = foreign key (FirstName LastName)
      references Master_List
      on delete restrict on update restrict;
   ic create Salary_Chk = check(where=(Salary le 90000));
run;
quit;
```

Below is a listing of these two data sets:

Figure 13.11: Listing of Data Sets Master_List and Salary

Listing of MASTER LIST

Obs	FirstName	LastName	DOB	Gender
1	Julie	Chen	07/07/1988	F
2	Nicholas	Schneider	04/15/1966	M
3	Joanne	DiMonte	06/15/1983	F
4	Roger	Clement	09/11/1988	M

Listing of SALARY

Obs	FirstName	LastName	Salary
1	Julie	Chen	54123
2	Nicholas	Schneider	56877
3	Joanne	DiMonte	67800
4	Roger	Clement	42000

If you run PROC CONTENTS on data set Master_List, you will see a listing of both the general and referential constraints. Below is a section of the PROC CONTENTS output showing this information:

Figure 13.12: Partial Listing of PROC CONTENTS for Data Set Master_List

#	Integrity Constraint	Type	Variables	Where Clause	Reference	On Delete	On Update
	Alphabetic List of Integrity Constraints						
1	Gender_Chk	Check		Gender in ('F', 'M')			
2	Prime_Key	Primary Key	FirstName LastName				
	Foreign_Key	Referential			WORK.SALARY	Restrict	Restrict

#	Index	Unique Option	Owned by IC	# of Unique Values	Variables
	Alphabetic List of Indexes and Attributes				
1	Prime_Key	YES	YES	4	FirstName LastName

You can see in this output that both your primary key constraint (Prime_key) and your foreign key constraint (Foreign_key) were created.

Attempting to Delete a Primary Key When a Foreign Key Still Exists

The next program demonstrates what happens when you attempt to delete an observation from a parent data set when that key still exists in a child data set. Program 13.15, without any referential constraints, would delete "Joanne DiMonte" from the Master_List data set. However, because of the referential constraint, no observations are deleted.

Program 13.15: Attempting to Delete a Primary Key When a Foreign Key Still Exists

```
*Attempt to delete an observation in the Master_List;
data Master_List;
   modify Master_List;
   if FirstName = 'Joanne' and LastName = 'DiMonte' then remove;
run;

title "Listing of Master_List";
proc print data=Master_List;
run;
```

Because of the referential constraint, the data set is not altered. Below is a copy of the SAS Log and a listing of the Master_List data set:

Figure 13.13: SAS Log After Running Program 13.15

```
60    data Master_List;
61       modify Master_List;
62       if FirstName = 'Joanne' and LastName = 'DiMonte' then remove;
63    run;

ERROR: Unable to delete observation. One or more foreign keys exist,
contain matching value(s), and are controlled by RESTRICT
referential action.
NOTE: The SAS System stopped processing this step because of errors.
NOTE: There were 3 observations read from the data set WORK.Master_List.
NOTE: The data set WORK.MASTER_LIST has been updated. There were 0
observations rewritten, 0 observations added and 0 observations
deleted.
NOTE: There were 0 rejected updates, 0 rejected adds, and 1 rejected
deletes.
```

Figure 13.14: Listing of Data Set Master_List

Listing of Master_List

Obs	FirstName	LastName	DOB	Gender
1	Julie	Chen	07/07/1988	F
2	Nicholas	Schneider	04/15/1966	M
3	Joanne	DiMonte	06/15/1983	F
4	Roger	Clement	09/11/1988	M

As you can see, no observations were deleted from the Master_List data set.

Attempting to Add a Name to the Child Data Set

This section demonstrates how a referential constraint prevents you from adding a foreign key value to the child data set when that key does not exist in the parent data set. To so this, we will attempt to add a new first and last name to the Salary data set that does not already exist in the Master_List data set. Take a look at the following:

Program 13.16: Attempting to Add a Name to the Child Data Set

```
data Add_Name;
   informat FirstName LastName $12. Salary comma10.;
   input FirstName LastName Salary;
   format Salary dollar9.;
datalines;
David York 77,777
;


proc append base=Salary data=Add_Name;
run;
```

In this program, you are using PROC APPEND to add observations from Add_Name to the end of the Salary data set.

The name "David York" does not exist in the Master_List data set. Therefore, the referential constraint should prevent you from adding this name to the Salary data set. Below is the SAS Log from running Program 13.16:

```
50   data Add_Name;
51      informat FirstName LastName $12. Salary comma10.;
52      input FirstName LastName Salary;
53      format Salary dollar9.;
54   datalines;

NOTE: The data set WORK.ADD_NAME has 1 observations and 3 variables.
NOTE: DATA statement used (Total process time):
      real time            0.01 seconds
      cpu time             0.01 seconds


56   ;
57
58   proc append base=Salary data=Add_Name;
59   run;

NOTE: Appending WORK.ADD_NAME to WORK.SALARY.
WARNING: Observation was not added/updated because a matching primary key
value was not found for foreign key Foreign_Key.
(Occurred 1 times.)
NOTE: There were 1 observations read from the data set WORK.ADD_NAME.
NOTE: 0 observations added.
NOTE: The data set WORK.SALARY has 4 observations and 3 variables.
```

Demonstrating How to Delete a Referential Constraint

If you want or need to delete a referential constraint, you can use PROC DATASETS (the same way you deleted general integrity constraints earlier). You also need to delete all foreign keys before you can delete a primary key. As an example, if you want to remove the referential constraints between the Master_List and the Salary data sets, you could proceed as follows:

Program 13.17: Demonstrating How to Delete a Referential Constraint

```
*delete prior referential integrity constraint;
*Note: Foreign key must be deleted before the primary key can be deleted;
proc datasets library=Work nolist;
   modify salary;
   ic delete Foreign_Key;
run;

   modify Master_List;
   ic delete Prime_Key;
run;
quit;
```

Notice that you remove the foreign key **before** the primary key since you cannot delete a primary key constraint unless all of the foreign key constraints have been deleted first.

Demonstrating the CASCADE Feature of a Referential Constraint

This example shows how updating primary key values will automatically update the corresponding foreign key values in all referencing foreign key data sets.

In the program that follows, realize that the two referential constraints were first deleted before being redefined. Program 13.18 creates the referential constraint that will change the value of FirstName and LastName in the Salary (child) data set when the value is changed in the Master_List (parent) data set:

Program 13.18: Demonstrating the CASCADE Feature of a Referential Integrity Constraint

```
proc datasets library=Work nolist;
   modify Master_List;
   ic create prime_key = primary key (FirstName LastName);
run;

   modify Salary;
   ic create foreign_key = foreign key (FirstName LastName)
      references Master_List
   on delete RESTRICT on update CASCADE;
run;
quit;

data Master_List;
   modify Master_List;
   if FirstName = 'Roger' and LastName = 'Clement' then
      LastName = 'Cody';
run;
```

```
title "Master List";
proc print data=Master_List;
run;

title "Salary";
proc print data=Salary;
run;
```

Because you have chosen CASCADE for the action to take place when you update a value of the primary key, the value of the two variables FirstName and LastName change in the Salary data set. Here are listings of both data sets:

Figure 13.15: Listing of Data Sets Master_List and Salary

Master List

Obs	FirstName	LastName	DOB	Gender
1	Julie	Chen	07/07/1988	F
2	Nicholas	Schneider	04/15/1966	M
3	Joanne	DiMonte	06/15/1983	F
4	Roger	Cody	09/11/1988	M

Salary

Obs	FirstName	LastName	Salary
1	Julie	Chen	54123
2	Nicholas	Schneider	56877
3	Joanne	DiMonte	67800
4	Roger	Cody	42000

Notice that Roger Clement has been changed to Roger Cody (my brother, by the way) in the Salary data set.

> In SAS documentation, it is recommended that you use the CASCADE feature of referential constraints with caution because it can be CPU intensive, especially with large data sets.

Demonstrating the SET NULL Feature of a Referential Constraint

You might want to delete a primary key in a parent data set and set all the foreign key references to a missing value (leaving all the other data for that observation intact). To accomplish this, you use the clause ON DELETE SET NULL, as demonstrated in Program 13.19 (Note: The two data sets Master_List and Salary were returned to their original condition):

Program 13.19: Demonstrating the SET NULL Feature of a Referential Constraint

```
proc datasets library=Work nolist;
   modify Master_List;
   ic create primary key (FirstName LastName);
run;

   modify Salary;
   ic create foreign key (FirstName LastName) references Master_List
   on delete SET NULL on update CASCADE;
run;
quit;
```

```
data Master_List;
   modify Master_List;
   if FirstName = 'Roger' and LastName = 'Clement' then
      remove;
run;

title "Master List";
proc print data=Master_List;
run;
title "Salary";
proc print data=Salary;
run;
```

When you run this program, the observation for Roger Clement is deleted and the values of FirstName and LastName are set to missing values (null) in the Salary data set (as shown below):

Figure 13.16: Listing of Data Sets Master_List and Salary

Master List

Obs	FirstName	LastName	DOB	Gender
1	Julie	Chen	07/07/1988	F
2	Nicholas	Schneider	04/15/1966	M
3	Joanne	DiMonte	06/15/1983	F

Salary

Obs	FirstName	LastName	Salary
1	Julie	Chen	54123
2	Nicholas	Schneider	56877
3	Joanne	DiMonte	67800
4			42000

Conclusions

Once you have spent time and energy cleaning a data set, it makes sense to add integrity constraints to the data set so that it is less likely for data errors to be added later. In most situations, you will want to create an audit trail data set as well. If you appended new data to an existing data set without an audit trail data set, you would not be able to tell which observations were added and which were not. Although this chapter briefly describes referential integrity constraints, they are complicated and you will most likely need further documentation and experimentation before using them.

Chapter 14: A Summary of Useful Data Cleaning Macros

Introduction

This book contains a number of macros that you can use to look for data errors. The purpose of this chapter is to collect, in one place, some of the macros that this author has found useful. Each macro will include a sample calling sequence.

A Macro to Test Regular Expressions

The macro %Test_Regex allows you to easily test a regular expression against a character string.

```
%macro Test_Regex(Regex=, /*Your regular expression*/
                  String= /*The string you want to test*/);
   data _null_;
      file print;
      put "Regular Expression is: &Regex " /
          "String is: &String";
      Position = prxmatch("&Regex","&String");
      if position then put "Match made starting in position " Position;
      else put "No match";
   run;
%Mend Test_Regex;
```

Here are some sample calling sequences:

```
%Test_Regex(Regex=/cat/,String=there is a cat there)
%Test_Regex(Regex=/([A-Za-z]\d){3}\b/, String=a1b2c3)
%Test_Regex(Regex=/([A-Za-z]\d){3}\b/, String=1a2b3c)
```

A Macro to List the *n* Highest and Lowest Values of a Variable

The %HighLow macro takes the *n* highest and lowest values for a numeric variable and lists these, along with an ID variable.

```
*Macro Name: HighLow
Purpose: To list the "n" highest and lowest values
Arguments: Dsn      - Data set name (one- or two-level)
           Var      - Variable to list
           IDvar    - ID variable
           N        - Number of values to list (default = 10)
example: %HighLow(Dsn=Clean.Patients,
                  Var=HR,
                  IDvar=Patno,
                  N=7)
;
%macro HighLow(Dsn=,     /* Data set name            */
              Var=,      /* Variable to list         */
              IDvar=,    /* ID Variable              */
              N=10       /* Number of high and low
                            values to list.
                            The default number is 10 */);

   proc sort data=&Dsn(keep=&IDvar &Var
                       where=(&Var is not missing))
                       out=Tmp;
      by &Var;
   run;

   data _null_;
      if 0 then set Tmp nobs=Number_of_Obs;
      High = Number_of_Obs - %eval(&N - 1);
      call symputx('High_Cutoff',High);
      stop;
   run;

   title "&N Highest and Lowest Values for &Var";
   data _null_;
   set Tmp(obs=&N)                     /* 'n' lowest values  */
      Tmp(firstobs=&High_Cutoff); /* 'n' highest values */
   file print;
   if _n_ le &N then do;
      if _n_ = 1 then put / "&N Lowest Values";
      put "Patno = " &IDvar @15 "Value = " &Var;
   end;
   else if _n_ ge %eval(&N + 1) then do;
      if _n_ = %eval(&N + 1) then put / "&N Highest Values";
```

```
        put "&IDvar = " &IDvar @15 "Value = " &Var;
    end;
    run;

    proc datasets library=work nolist;
        delete Tmp;
    run;
    quit;
%mend HighLow;
```

Here are two sample calling sequences:

The following macro call will list the seven highest and lowest values for HR in the data set Clean.Patients, using the variable Patno as the ID variable.

```
%HighLow(Dsn=Clean.Patients,
         Var=HR,
         IDvar=Patno,
         N=7)
```

The next macro call will list the 10 highest and lowest values of Deposit in data set Banking, using the variable Account as the ID variable. Ten values are listed because, without specifying a value for *n*, a default value of 10 is used.

```
%HighLow(Dsn=Banking,
         Var=Deposit
         IDvar=Account)
```

A Macro to List the *n*% Highest and Lowest Values of a Variable

This macro is similar to the previous (%HighLow) macro, except that you specify the number of values to print as a percentage of the total number of observations in the data set.

```
*------------------------------------------------------------------*
| Program Name: HighLowPcnt.sas                                    |
| Purpose: To list the n percent highest and lowest values for     |
|          a selected variable.                                    |
| Arguments: Dsn     - Data set name                               |
|            Var     - Numeric variable to test                    |
|            Percent - Upper and Lower percentile cutoff           |
|            Idvar   - ID variable to print in the report          |
| Example: %HighLowPcnt(Dsn=clean.patients,                        |
|                       Var=SBP,                                   |
|                       Percent=5,                                 |
|                       Idvar=Patno)                               |
*------------------------------------------------------------------*;

%macro HighLowPcnt(Dsn=,   /* Data set name                      */
               Var=,       /* Variable to test                   */
               Percent=,   /* Upper and lower percentile cutoff  */
               Idvar=      /* ID variable                        */);
```

```
   ***Compute upper percentile cutoff;
   %let Upper = %eval(100 - &Percent);

   proc univariate data=&Dsn noprint;
      var &Var;
      id &Idvar;
      output out=Tmp pctlpts=&Percent &Upper pctlpre = Percent_;
   run;

   data HiLow;
      set &Dsn(keep=&Idvar &Var);
      if _n_ = 1 then set Tmp;
      if &Var le Percent_&Percent and not missing(&Var) then do;
         range = 'Low ';
         output;
      end;
      else if &Var ge Percent_&Upper then do;
         range = 'High';
         output;
      end;
   run;

   proc sort data=HiLow;
      by &Var;
   run;

   title "Highest and Lowest &Percent% for Variable &var";
   proc print data=HiLow;
      id &Idvar;
      var Range &Var;
   run;

   proc datasets library=work nolist;
     delete Tmp HiLow;
   run;
   quit;

%mend HighLowPcnt;
```

The following macro call will list the top and bottom 10 percent for the variable Deposit in the Banking data set, using Account as the ID variable.

```
%HighLowPcnt(Dsn=Clean.Banking,
             Var=Deposit,
             Percent=10,
             Idvar=Account)
```

A Macro to Perform Range Checks on Several Variables

The two macros, %Errors and %Report, allow you to specify valid ranges for a list of numeric variables. In addition, you can elect to ignore missing values (the default) or to flag missing values as errors (Missing=error).

```
*Program Name: Errors.Sas
 Purpose: Accumulates errors for numeric variables in a SAS
         data set for later reporting/
         This macro can be called several times with a
         different variable each time. The resulting errors
         are accumulated in a temporary SAS data set called
         errors.

*Macro variables Dsn and IDvar are set with %Let statements before
 the macro is run;

%macro Errors(Var=,     /* Variable to test     */
              Low=,     /* Low value            */
              High=,    /* High value           */
              Missing=ignore
                        /* How to treat missing values        */
                        /* Ignore is the default. To flag     */
                        /* missing values as errors set       */
                        /* Missing=error                      */);

data Tmp;
   set &Dsn(keep=&Idvar &Var);
   length Reason $ 10 Variable $ 32;
   Variable = "&Var";
   Value = &Var;
   if &Var lt &Low and not missing(&Var) then do;
      Reason='Low';
      output;
   end;
   %if %upcase(&Missing) ne IGNORE %then %do;
   else if missing(&Var) then do;
      Reason='Missing';
      output;
   end;
   %end;

   else if &Var gt &High then do;
       Reason='High';
      output;
      end;
      drop &Var;
   run;

   proc append base=errors data=Tmp;
   run;

%mend errors;
```

```
%macro report;
   proc sort data=Errors;
      by &Idvar;
   run;

   proc print data=errors;
   title "Error Report for Data Set &Dsn";
      id &Idvar;
      var Variable Value Reason;
   run;

   proc datasets library=work nolist;
      delete errors;
      delete tmp;
   run;
   quit;

%mend report;
```

It is important to remember to submit two %LET statements to specify the data set name and the name of the
ID variable before running the %Errors macro. Here is an example:

In this example, missing values will be reported for HR but ignored for SBP and DBP. Because the macro
variable Missing is not referenced in the test for DBP ranges, the default action of IGNORE is used. You
might want to include the instruction Missing=Ignore, even though it is the default, because people not
familiar with the macro (or yourself if you haven't used it in a while) might not remember the default action.

```
***Set two macro variables;
%let Dsn=Clean.Patients;
%let IDvar = Patno;

%Errors(Var=HR, Low=40, High=100, Missing=error)
%Errors(Var=SBP, Low=50, High=240, Missing=ignore)
%Errors(Var=DBP, Low=35, High=130)

***Generate the report;
%report
```

A Macro that Uses Trimmed Statistics to Automatically Search for Outliers

The %Auto_Outliers macro uses trimmed statistics to list possible data errors in numeric data. Deciding how
much to trim should be based on your knowledge of the distribution of the variables you are investigating. A
default value of a 10% trim is coded into the macros, and the cutoff value for possible outliers is set to a
default value of 2. You will probably want to increase this latter value for large data sets.

```
*Method using automatic outlier detection;
%macro Auto_Outliers(
   Dsn=,      /* Data set name                          */
   ID=,       /* Name of ID variable                    */
```

```
    Var_list=, /* List of variables to check            */
               /* separate names with spaces            */
    Trim=.1,   /* Integer 0 to n = number to trim       */
               /* from each tail; if between 0 and .5,  */
               /* proportion to trim in each tail       */
    N_sd=2     /* Number of standard deviations         */);

    ods listing close;
    ods output TrimmedMeans=Trimmed(keep=VarName Mean Stdmean DF);
    proc univariate data=&Dsn trim=&Trim;
       var &Var_list;
    run;
    ods output close;

    data Restructure;
        set &Dsn;
        length VarName $ 32;
        array Vars[*] &Var_list;
        do i = 1 to dim(Vars);
            VarName = vname(Vars[i]);
            Value = Vars[i];
            output;
        end;
        keep &ID VarName Value;
    run;

    proc sort data=Trimmed;
       by VarName;
    run;

    proc sort data=restructure;
       by VarName;
    run;

    data Outliers;
       merge Restructure Trimmed;
       by VarName;
       Std = StdMean*sqrt(DF + 1);
       if Value lt Mean - &N_sd*Std and not missing(Value)
          then do;
              Reason = 'Low  ';
              output;
          end;
       else if Value gt Mean + &N_sd*Std
          then do;
          Reason = 'High';
          output;
       end;
    run;

    proc sort data=Outliers;
       by &ID;
    run;
```

```
    ods listing;
    title "Outliers Based on Trimmed Statistics";
    proc print data=Outliers;
        id &ID;
        var VarName Value Reason;
    run;

    proc datasets nolist library=work;
        delete Trimmed;
        delete Restructure;
    run;
    quit;
%mend Auto_Outliers;
```

It is important to remember that the variables in the Var_List are separated by blanks. You might want to run this macro for groups of similar variables where you plan to use the same trim and number of standard deviations.

On a personal note, this author recently used this macro on a data set consisting of such variables as Age and BMI (body mass index) along with several other related variables, and it was very successful in identifying errors. The majority of errors in this data set resulted from switching the order of digits. For example, an age of 16 was entered, but the person was 61.

A sample call of this macro is shown next:

```
%Auto_Outliers(Dsn=Clean.Patients,
               Id=Patno,
               Var_List=HR SBP DBP,
               Trim=.1,
               N_Sd=2)
```

A Macro to Search a Data Set for Specific Values Such as 999

Many data sets include special values such as 999 or 888 to indicate a special condition or, most likely, a missing value. The macro %Find_Value will automatically search all the numeric variables in a SAS data set for a specified value and present you with a report of which variables used the special value and how many times they used it.

```
*Macro name: Find_Value.sas
Purpose: Identifies any specified value for all numeric variables
Calling arguments: dsn=    sas data set name
                   value= numeric value to search for
Example:  To find variable values of 999 in data set Test, use
          %Find_Value(dsn=Test, Value=999);
%macro Find_Value(Dsn,  /* The data set name */
                  Value= /* Value to look for */ );

    title "Variables with &Value as Missing Values in Data Set &Dsn";
    data Tmp;
        set &Dsn;
        file print;
```

```
      length Varname $ 32;
      array Nums[*] _numeric_;
      do iii = 1 to dim(Nums);
         if Nums[iii] = &Value then do;
         Varname = vname(Nums[iii]);
         output;
         end;
      end;
      keep Varname;
   run;

   proc freq data=Tmp;
      tables Varname / out=Summary(keep=Varname Count)
                       nocum;
   run;

   proc datasets library=Work nolist;
      delete Tmp;
   run;
   quit;
%mend Find_Value;
```

The following macro call will list all the numeric variables in data set Test with a value of 999.

```
%Find_Value(dsn=Test, Value=999)
```

A Macro to Check for ID Values in Multiple Data Sets

The %Check_ID macro will check if an ID variable is missing from any number of SAS data sets.

```
*Program Name: Check_ID.sas
 Purpose: Macro which checks if an ID exists in each of n files
 Arguments: The name of the ID variable, followed by as many
            data sets names as desired, separated by BLANKS
 Example: %Check_ID(ID = Patno,
                Dsn_list=One Two Three);

%macro Check_ID(ID=,        /* ID variable               */
               Dsn_list=  /* List of data set names,   */
                          /* separated by spaces       */);
   %do i = 1 %to 99;
     /* break up list into data set names */
     %let Dsn = %scan(&Dsn_list,&i,' ');
     %if &Dsn ne %then %do; /* If non null data set name   */
        %let n = &i;         /* When you leave the loop, n will */
                             /* be the number of data sets    */
        proc sort data=&Dsn(keep=&ID) out=Tmp&i;
           by &ID;
        run;
     %end;
   %end;
```

```
   title  "Report of data sets with missing ID's";
data _null_;
   file print;
   merge

   %do i = 1 %to &n;
      Tmp&i(in=In_Tmp&i)
   %end;

   end=Last;
   by &ID;

   if Last and n eq 0 then do;
      put "All ID's Match in All Files";
      stop;
   end;

   %do i = 1 %to &n;
      %let Dsn = %scan(&Dsn_list,&i);
      if not In_Tmp&i then do;
         put "ID " &ID "missing from data set &dsn";
         n + 1;
      end;
   %end;

   run;
%mend Check_ID;
```

The macro call below will report if values of the variable Patno are missing from data sets One, Two, or Three.

```
%check_ID(ID=Patno, Dsn_List=One Two Three)
```

Conclusions

You can download all of these macros (and all the programs and data sets used in this book) from my author web site:

```
support.sas.com/cody
```

You might want to place the macros in your macro library and use the SAS AUTOCALL facility to make the macros available at any time, without having to submit an %INCLUDE statement.

Index

Ready to take your SAS® and JMP®skills up a notch?

Be among the first to know about new books,
special events, and exclusive discounts.
support.sas.com/newbooks

Share your expertise. Write a book with SAS.
support.sas.com/publish

sas.com/books
for additional books and resources.

sas

THE POWER TO KNOW®

SAS and all other SAS Institute Inc. product or service names are registered trademarks or trademarks of SAS Institute Inc. in the USA and other countries. ® indicates USA registration.
Other brand and product names are trademarks of their respective companies. © 2017 SAS Institute Inc. All rights reserved. M1588359 US 0217

7

CPSIA information can be obtained
at www.ICGtesting.com
Printed in the USA
BVOW08s0528041017
496641BV00005B/31/P